Digital Methods for
Signal Analysis

Other works by Dr Beauchamp

Signal Processing, George Allen & Unwin, 1973
Walsh Functions and their Applications, Academic Press, 1975
Exploitation of Seismograph Networks (ed.), Noordhoff Press, 1975
Interlinking of Computer Networks (ed.), D. Reidel, 1979

Digital Methods for Signal Analysis

by

K. G. BEAUCHAMP, C.G.I.A., C.Eng., Ph.D., M.I.E.E.
Director, Computer Services,
University of Lancaster

and

C. K. YUEN, B.Sc., M.Sc., Ph.D.
Lecturer, Information Science Department,
University of Tasmania

London
GEORGE ALLEN & UNWIN
Boston   Sydney

First published 1979

GEORGE ALLEN & UNWIN LTD
40 Museum Street, London WC1A 1LU

© K. G. Beauchamp and C. K. Yuen, 1979

British Library Cataloguing in Publication Data

Beauchamp, Kenneth George
    Digital methods for signal analysis.
    1. Signal processing
    I. Title   II. Yuen, C K
    621.38'043'02854044   TK5102.5   78–41171

    ISBN 0–04–621027–X

Typeset in 10 on 12 point Times by
George Over Limited, London and Rugby
and printed in Great Britain
by University Press, Cambridge

CONTENTS

Preface

This book first appeared in 1973 under the title *Signal Processing, using analog and digital techniques*. Since then the subject has undergone considerable change, enough, in fact, to call for a second, revised edition under a new title.

Substantial reductions in the cost of digital computation have occurred, accompanied by an improvement in performance, speed, memory capacity and ease of programming. Much progress has also been made in the design of special-purpose data acquisition and on-line processing systems. Whereas previously analog methods could offer benefits of simplicity and low cost against the wide applicability and numerical accuracy of digital techniques, time has established clearly the superiority of the latter in almost all signal processing applications.

In consequence, this edition, as the title implies, is concerned solely with digital methods. Recent developments in non-sinusoidal processing methods, appropriate mainly to digital exploitation, are included and a chapter is devoted to the processing of pictorial information. Progress in this latter field has been concomitant with the advance of digital computers themselves, permitting the mathematical manipulation of large matrix representations of images for such purposes as feature extraction, picture enhancement and improvement, bandwidth compression and many others.

Readers of the earlier edition will find in this volume a shift in emphasis which we hope will render the material more readily understood by a wider audience. In the early years, signal processing applications were to be found mostly in engineering and the physical sciences. Now the tools of signal processing are employed also by investigators in such diverse fields as biology, medicine, economics and the social sciences. The authors consider that these newer applications require not new material or new analysis methods, but a change of presentation to make the existing material more comprehensible to those not having an engineering background. We have tried to achieve this by the addition of explanatory material on the mathematical and statistical basis of signal analysis, together with the identification of those subjects which are considered to be more specialised and which may be omitted on first reading. These sections have been marked with an asterisk *.

The first three chapters provide basic introductory and mathematical concepts appropriate to the subject. Chapter 1 outlines the scope of the material which follows and describes some of the practical and mathematical problems that are found in the analysis of discrete and digitised information. Chapter 2 looks at the statistical nature of the signal processing operation whilst Chapter 3 is concerned with data transformation into other more convenient domains.

This basic information is applied in the remaining chapters to those processing operations on digital data designed to enhance their information content or to extract meaningful characteristics such as spectral composition and correlation information. Chapter 4 is devoted to spectral analysis and includes a brief statement of non-sinusoidal techniques. Chapter 5 deals with correlation analysis and associated methods of signal enhancement. Chapter 6 presents a general introduction to digital filtering techniques. Finally Chapter 7 is concerned entirely with the processing of two-dimensional data. The subject is considered too extensive for any complete discussion and we have only selected several topics which we feel fit in best with earlier chapters.

It must be emphasised that this book is directed primarily towards those who are confronted with practical processing problems and understanding at the 'grass roots' level. Whilst the theory of signal analysis is complex and highly mathematical, involving such concepts as stochastic processes and Fourier integrals, the actual computing procedures: spectral estimation, correlation, filtering etc., are often quite simple. One can learn the computing procedures without any of the mathematical theory, but as Richard Hamming says, **the purpose of computing is insight, not numbers**. Without some grasp of the basic theory, it would be difficult to understand the meaning of the computed results. On the other hand, to try and provide a complete and rigorous discussion of the theory of signal analysis would make the book far too specialised. It is our hope that we have maintained a correct balance between the two aspects, and to have explained the basic concepts sufficiently clearly to form a starting point from which the reader may explore on his own and learn by experience. The authors would welcome the readers' comments on the book. In particular, discussions specifically related to Chapters 1, 3, 6 and 7 should be directed to the first author; those related to the other chapters to the second author.

The authors would like to acknowledge with grateful thanks the assistance given to them by very many people, including Professor V. Cappellini of the Consiglio Nazionale Delle Ricerche, Dr T. M. Cannon of the University of Utah, Dr O. L. Frost of Argo Systems Inc., California, Professor H. F. Harmuth of the Catholic University of America, Professor T. S. Huang of Purdue University, Dr J. O. Thomas of Imperial College, Dr A. G. Piersol of the University of Southern California, and Dr D. Fraser of CSIRO, Australia. Finally, we would like to express our appreciation to Mrs P. Taylor and several other typists for their work in the preparation of the many manuscripts.

K. G. Beauchamp
University of Lancaster
C. K. Yuen
University of Tasmania

List of Symbols

a, b, A, B,	Constants
a_k, b_k	Fourier coefficients
A_k, B_k	Complex Fourier coefficients
$\mathbf{A_k}$	Digital picture element (pixel)
$\mathbf{A}$	Matrix
$\mathbf{A^{-1}}$	Inverse matrix
b_i	Bit position within a digital word
B	Bandwidth
CAL(k,t)	CAL function series
C_k	Modulus of Fourier series
C(x,y)	Covariance of x and y
$C_{xy}(f)$	Co-spectral density function
exp	Exponential, $2 \cdot 71828$
f	Frequency (Hz)
f_c	Filter cut-off frequency
f_n	Nyquist frequency
f_s	Sampling frequency or sampling rate
f(x,y)	Two-dimensional image
F(x)	Function of x
g(x,y)	Two-dimensional image
$\mathbf{G_k}$	Filter weight matrix
h	Sampling interval
h_k	Filter weight series
h(t)	Impulse response function
h(u,v,x,y)	Point spread function
HAR(n,t), HAR(n,m,t)	Haar function series
H(s)	Laplace transform function of h(t)
H(ω)	Fourier transform of h(t)
H(m,n)	Transformed point spread function
Hz	Frequency unit
H(z)	Z-transform function
i, k, m, n	Series coefficients
$\mathbf{I}$	Identity operator
Im	Imaginary part of
j	$\sqrt{-1}$
J	Bessel function
K	Constant, filter gain constant
$K_x(\tau)$	Cepstrum function

$K(u,v)$	Two-dimensional cepstrum function
$\log, \log_{10}$	Logarithm to base 10
$\log_e, \log_n$	Logarithm to base e, natural logarithm
n	Degrees of freedom
n!	Factorial n
$n(t)$	Noise function
N	Number of terms in a series
$O(x,y)$	Two-dimensional object image
$p(x)$	Probability density function
$p(x,y)$	Joint probability density function
P	Probability
$P(x)$	Probability distribution function
$P(x,y)$	Joint probability distribution function
$P_k(\omega), P(k)$	Periodogram function
$Q_{xy}(\omega)$	Quadrature spectral density function
r_s	Signal-to-noise ratio
$r_x(\tau)$	Normalised auto-correlation function
Re	Real part of
$Rd(i,t)$	Rademacher function series
$R(o)$	Auto-correlation function for zero lag
$R(\tau), R_{xx}(\tau), R(t,s)$	Auto-correlation function for x
$R_{xy}(\tau)$	Cross-correlation function between x and y
$R_x^c(\tau)$	Circular correlation function
$s = j\omega$	Laplace frequency transformation
$\text{sinc}(t)$	$\dfrac{\sin(\pi t)}{\pi t}$
S	Standard deviation for a sample
$\hat{S}$	Estimated value of S
$SAL(k,t)$	SAL function series
$S_x(\omega_1,\omega_2)$	Generalised spectral density function
$S_x(\omega)$	Power spectral density function
$S_{xy}(\omega)$	Cross spectral density function
t_n	Time series
T	Record time, record length, time, transmission
$V(x)$ or σ_x^2	Variance of x
$W(t)$	Weighting function (time window)
$W(f)$	Window function (frequency window)
$WAL(n,t)$	Walsh function series
$W_f(\phi)$	Weighting function
$\bar{x}$	Average value of x(t)
$<x>$	Ensemble average value
x_i	Discrete series
$x(t)$	Continuous time series
$\overline{x(t)}$	Time average of x(t)
$<x(t)>$	Ensemble average of x(t)
$\overline{x^2}$	Mean-square value of x
$x_{i,k}$	Two-dimensional discrete series

x^*	Complex conjugate of x
$XC(f)$	Discrete CAL transform
$Xc(f)$	Fourier cosine transform
$X(f)$	Fourier transform of x
X_k, X_n	Discrete transform of x
$X_{n,m}$	Discrete two-dimensional transform
$X_s(f)$	Fourier sine transform
$Xs(f)$	Discrete SAL transform
$y(t)$	Time series, output series
$z = \exp(sT)$	Z-transform coefficient
$z^{-1} = \exp(-sT)$	Inverse Z-transform coefficient

Greek Symbols

α, β, γ	Constants
$\beta(\hat{x})$	Bias of estimate $\hat{x}$
$\Gamma(\ )$	Gamma function
$\delta(t)$	Delta function, Dirac impulse function
δx	Increment of x
Δ	Increment
ϵ	Normalised standard error, estimation error
θ	Phase angle, angle of rotation
κ_{xy}	Coherency spectrum
$\mu, <\ >$	Ensemble average
ξ	Random variable
π	Product
ρ	Correlation coefficient
σ	Standard deviation
σ^2	Variance
τ	Time delay lag
ϕ	Phase angle
χ^2_n	Chi-squared variable having n degrees of freedom
$\omega = 2\pi f$	Angular frequency
ω_c	Cut-off angular frequency
ω_0	Fundamental angular frequency

Other Symbols

$<z$	angle z
$*$	Convolution, complex conjugate
$\oplus$	Modulo-2 addition
$!$	Factorial
∞	Infinity
$\propto$	Proportional to
$\rightarrow a$	Approaches value a
$\Rightarrow A$	Transforms to A

Chapter 1

Introductory

1.1   What is a Signal?

We can recognise a large number of sources providing information on the presence or change of a physical phenomenon. If arrangements are made to convert this information from its original physical form (e.g. temperature, pressure, vibration etc.), into a directly related electrical quantity (e.g. potential, current, power, etc.), then this electrical quantity is called a **signal**. This signal may merely indicate the presence of the physical phenomenon or it may provide one or more related parameters concerning its behaviour. Many signals behave as continuous processes during the period of their acquisition and provide a history of the quantity being measured. This continuing but finite length record of the process is termed a **time-history**.

In the material that follows, the most common independent variable is time, and we can express the time-history as either a continuous function of time, $x(t)$, or a discrete sampled function, x_i for a single dependent variable ($i=0,1,\ldots n$).

The second most common independent variable is frequency for which the series of values obtained may be expressed in terms of a direct function of frequency, $x(f)$, and $x_i(f)$ or as a function of angular frequency, $x(\omega)$, $x_i(\omega)$, or as a complex frequency, $x(j\omega)$, $x_i(j\omega)$.

Many other independent variables for the signal are possible. Examples are found in astronomical and economic data which may be related to spatial position or some item number identifying a specific object in a collection of similar objects for which the time of acquisition is irrelevant.

The mathematical analysis techniques to be described however are perfectly general and the substitution of one of these other independent variables for time is frequently carried out and requires only a rearrangement of terms.

The acquired signal may not always conveniently be obtained as a function of a single variable, although it may very often be considered as such. A common example is a signal which is both a function of frequency and time and expressed as, $x(f, t)$ or $x(\omega, t)$. Representation of the two-dimensional time-frequency history as a single equivalent independent variable or sequential set of variables is often attempted as part of the analysis procedure.

All signals acquired from contact with a physical environment can be expressed in the form of a time-history even if the independent variable is other than time. Before any attempt can be made to extract quantitative data from this, some information must be obtained concerning the environment from which the signal was derived, together with information concerning the signal itself, in order to permit recognition of the type of signal we are dealing with. Without this preliminary classification, meaningful analysis is not possible.

1.2   Signal Classification

Where the physical quantity being measured can be described explicitly in terms of mathematical relationships then it is likely that we can regard the derived signal in the same way. This type of signal is classed as **deterministic**, and may be identified further as belonging to one of a number of the groups discussed below. More generally the signal is likely to be **random**, that is, we will not be able to predict precisely its value at any given future instant in time. Analysis must then proceed in terms of statistical values from which it is possible that a deterministic relationship may be obtained. One criterion used to decide whether the signal is deterministic or random is to compare several sets of data obtained under identical conditions over a reasonable period of time. If similar results are obtained then the data are likely to be deterministic.

It is important to distinguish between the required information concerning the physical measurement and the distortion and extraneous noise that may be included with the available signal. A signal may appear to be random and yet yield deterministic data by applying a suitable form of analysis, e.g. auto-correlation. In such a case random analysis methods will be succeeded by deterministic measures later in the analysis procedure. Alternatively random data may themselves represent the phenomenon under investigation and the signal obtained will contain these random data mixed with periodic data of no interest. Analysis methods arranged to detect and remove the periodic component will then precede the required random signal analysis.

1.2.1   DETERMINISTIC SIGNALS

The mathematical form of the physical quantities represented by the deterministic signal, if known (or assumed), will either continuously repeat at regular intervals or decay to a zero value after a finite length of time. The former are known as **periodic signals** and the latter as **transients**.

Periodic signals can be considered as comprising one or more sinusoidal signals having an integral relationship with the period of repetition. This period, T, is defined as the interval of time over which the signal repeats itself. (We are considering only single-dimensional time-histories here.)

The simplest example is the sinusoidal function (see Fig. 1.1)

$$X(t) = A \sin(\omega_0 t + \theta) \qquad (1.1)$$

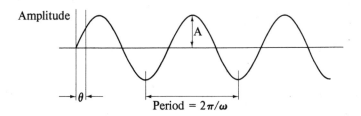

Fig. 1.1   A sinusoidal function

where A is a constant representing the peak amplitude of the wave-form and ω_0 is the angular frequency $= 2\pi f_0$. f_0 is the number of repetitions per unit time, or in technical terms, the cyclical frequency in Hz, and θ is the initial **phase angle** with respect to the time origin in radians.

The period of repetition, T, will be related to f_0 by

$$T = 1/f_0 \qquad (1.2)$$

A more general relationship for periodic signals is obtained by considering x(t) to comprise harmonically related sinusoids and expressed as a Fourier series, where each harmonic component repeats itself exactly for all values of t, i.e.

$$x(t) = A_0 + A_1 \sin(\omega_0 t + \theta_1) + A_2 \sin(2\omega_0 t + \theta_2) + \ldots A_n \cdot \sin(n\omega_0 t + \theta_n) \qquad (1.3)$$

or

$$x(t) = A_0 + \sum_{n=1}^{n=\infty} A_n \cdot \sin(n \cdot \omega_0 t + \theta_n) \ (n = 1, 2, 3, \ldots) \qquad (1.4)$$

where A_0 = mean level of signal, A_n = peak amplitude of the nth harmonic and θ_n = phase of the nth harmonic. The sinusoidal case, equation (1.1), will be seen to be a special case of equation (1.4) for n=1 and A_0=O. Several examples of non-sinusoidal periodic signals are illustrated in Fig. 1.2. The frequency content of such periodic signals may be expressed by **line spectra** as shown in Fig. 1.3. Individual frequencies comprising the signal are discrete and located at precise positions on the frequency axis.

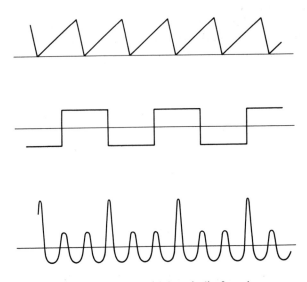

Fig. 1.2 Non-sinusoidal periodic functions

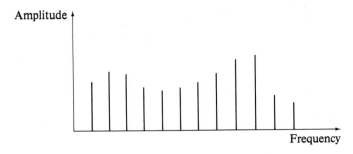

Fig. 1.3 Line spectra

 A deterministic signal may consist of the synthesis of several sinusoidal elements, which are not necessarily harmonically related. Such signals are common in communication engineering where heterodyne effects between two or more wave-forms take place producing a multi-frequency resultant having no precise repetition period. This combination may be produced by summing or multiplication of sinusoids and any combination of these. An example of this type of signal would be

$$x(t) = A_1 \sin \omega_1 t + A_2 \cos \omega_2 t + A_3 (\sin \omega_1 t)(\cos \omega_2 t) \qquad (1.5)$$

This will not result in a periodic signal but will exhibit similar spectral characteristics.

Transient signals are time-varying signals which reduce to zero value over a finite time interval. Some examples of these are given in Fig. 1.4.

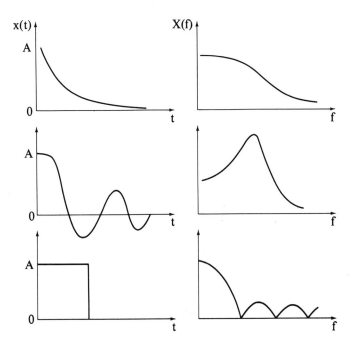

Fig. 1.4 Transient signals

Such signals are characterised by **continuous spectra** rather than line spectra, since in theory an infinite number of frequency components are present in the signal. Fourier series representation can no longer be used and spectral representation is obtained by means of the **Fourier integral**

$$X(f) = \int_{-\infty}^{\infty} x(t) \exp(-j2\pi ft) \, dt \qquad (1.6)$$

This is a complex quantity and can only be represented completely by an amplitude and phase value for $X(f)$ over the frequency range.

A modulus form $|X(f)|$ is often used to represent the spectra for a transient signal. An example is given in Fig. 1.5.

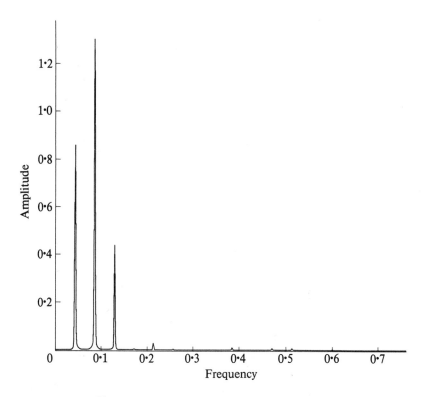

Fig. 1.5   Continuous spectral representation

1.2.2   RANDOM SIGNALS

With deterministic signals their functions can be obtained from the time-history, and generally made to repeat exactly under identical conditions. This is not the case with random signals where their functions cannot be determined explicitly and where probabilistic and statistical descriptions will need to be used.

This will be understood from the consideration of a particular random process which we will assume to produce a set of time-histories, known as an **ensemble**, each referenced to an identical time instant or commencement time, t = 0 (see Fig. 1.6). This could represent for example, an experiment producing random data, which is repeated N times to give an ensemble of N separate records. Statistical values for this ensemble can be obtained by considering the values of these records taken at specific instants in time.

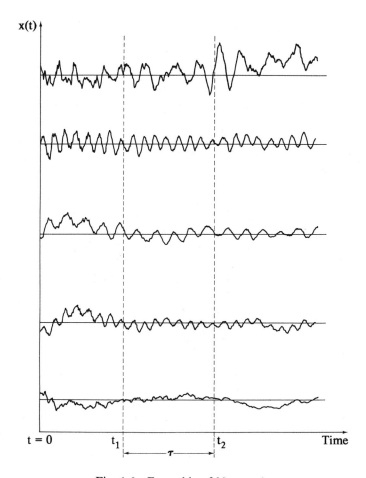

Fig. 1.6   Ensemble of N records

For example the average value existing at time, t_1 summed over the ensemble is

$$<x(t_1)> = \lim_{N \to \infty} \frac{1}{N} \sum_{k=1}^{N} x_k(t_1) \qquad (1.7)$$

The average value of the products of two samples taken at two separate times, t_1 and t_2, for each separate record (also known as the **auto-correlation function**) is

$$R(\tau) = \lim_{N \to \infty} \frac{1}{N} \sum_{k=1}^{N} x_k(t_1) \cdot x_k(t_2) \qquad (1.8)$$

$$(\tau = t_2 - t_1)$$

The process of arriving at these values is known as **ensemble averaging** and may be continued over the entire record length to provide a statistical description of the complex set of records.

A signal is defined as **stationary** if the values of $<x(t_1)>$ and $R(\tau)$ are found to remain constant for all possible values of t_1 and that $R(\tau)$ is dependent only upon the time displacement (τ). In a more realistic and practical case we say that the signal is stationary if $<x(t_1)>$ and $R(\tau)$ are constant over the finite record length, T. If the values of $<x(t_1)>$ and $R(\tau)$ vary with time, then the signal is defined as being **non-stationary**. This is the more general case although for very many practical situations the change with time is so slow as to permit broad classification as a stationary process. Under certain circumstances we can regard a signal as stationary by considering the statistical characteristics of a single long record. Referring to Fig. 1.7, if the record is partitioned into L equal length sections of length T, then the average value of any section M will be

$$\bar{x}_M = \lim_{T \to \infty} \frac{1}{T} \int_0^T x_M(t) \cdot dt \qquad (1.9)$$

Here the symbol $\bar{x}$ is used to denote a single time-history average to distinguish this from $<x>$ which represents an ensemble average.

The auto-correlation function taken over any section M will be

$$R_M(T) = \lim_{T \to \infty} \frac{1}{T} \int_0^T x_M(t) x \cdot (t+\tau) \cdot dt \qquad (1.10)$$

If this single record averaging process is also carried out for a given record taken from the ensemble of Fig. 1.6 and we find that

$$\bar{x}_M = <x>$$

and

$$R_M(\tau) = R(\tau) \qquad (1.11)$$

for all values of t and all sections L, then the process is called an **ergodic random process**. By definition this must also be a stationary process. This particular form of a stationary process has the property that all of its statistical quantities can be obtained from measurements carried out on a

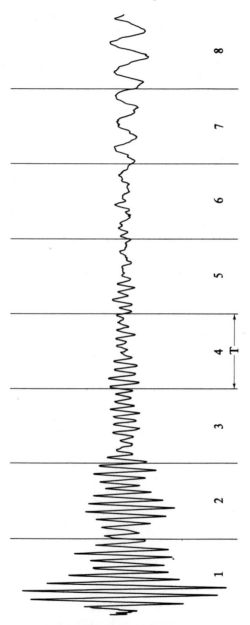

Fig. 1.7 Partitioning of a single long record

single record length. Often a process may be assumed to be ergodic and since many physical processes yield only a single realisation this ergodic assumption is an important one.

Whilst the assumption of stationarity is a necessary one for ergodic processes, a stationary random process need not be ergodic. We could, for example, have a stationary ensemble where each individual record expresses a signal at a different mean level so that the mean level of segments taken from each record would not be equal, although constant over the record length. In many practical circumstances non-stationarity may be regarded relative to some initial starting point, so that all subsequent records obtained from the same process will be assumed to have similar form. Thus a type of ergodic hypothesis may be applied which allows a single record to be examined and statistical conclusions drawn using only one realisation of the process. This is found to be an acceptable situation for slowly varying time-dependent signals although it is not to be considered as replacing the more rigorous analysis of an ensemble of records.

1.2.3 IDENTIFICATION REQUIREMENTS

Classification is dependent on prior knowledge concerning the signal itself. To assist signal identification it is necessary to have some idea of its domain limitations, e.g. frequency range, amplitude or power range and time duration. Since this information (other than time) is provided as an electrical analogy of the physical quantity being measured, details of the signal environment are also required so that quantitative conclusions can be drawn. We thus need to study the conversion characteristics of the transducers used, their accuracy and limitations. We will also need to make use of information gained from previous measurements or analytical examination. The signal may not be analysed in real-time and if a temporary storage media is used then its characteristics and time-translation property, if applicable, will also need to be known.

Finally, when considering the actual results obtained we have to realise that all measurements on random data are estimates and will be obtained at an accuracy dependent on the amount of data made available. Conclusions drawn from a single record should therefore be regarded as tentative and must be accompanied by a statement of the conditions of acquisition and measurement if gross subjective errors are to be avoided.

1.3   The Nature of Signal Analysis

Forms of analysis for random data rely heavily on statistics and probability for reasons stated earlier, and will be considered in detail in the succeeding chapters. The broad areas of analysis are found in the frequency, time, amplitude and power domains. Translation of information from one domain to another is frequently required to facilitate extraction of meaningful characteristics.

One of the earliest examples of analysis in the frequency domain was the use of the **periodogram** to determine 'hidden periodicities' in a time-history series. The periodogram was suggested by Schuster towards the end of the last century [1], and used to extract information concerning the variable nature of the earth's magnetic field. In its discrete form the periodogram is given by the function

$$P_k(\omega) = \frac{1}{N}\left[\left(\sum_{i=1}^{N} x_i \cdot \cos i \, \frac{2\pi}{\omega}\right)^2 + \left(\sum_{i=1}^{N} x_i \cdot \sin i \, \frac{2\pi}{\omega}\right)^2\right]$$

$$(i,k=1,2,\ldots,N) \tag{1.12}$$

The application of this function to the discrete data series, x_i, will identify the periodic nature of x_i by giving a peak value of $P_k(\omega)$ at $\omega = \omega_0$ and smaller peaks at other harmonic frequencies.

The essential idea behind the early use of the periodogram for certain time series was that if the periodic values could be determined accurately then sinusoidal wave-forms corresponding to these frequencies could be subtracted in sequence from the original signal leaving ultimately only the random independent series, e(t). This assumed that a model for the data series would take the form

$$x(t) = \sum_{i=1}^{N} A_i \cdot \sin(\omega_i t + B_i) + e(t) \tag{1.13}$$

where A_i and B_i are constants.

Whilst it has been successfully applied in this form to a few favourable data series [2] it may be shown to be productive of a number of spurious side-peaks which can be misleading [3], particularly as they will be found to extend over a considerable frequency range. Later workers found ways of avoiding this difficulty and under certain circumstances the periodogram can be applied successfully to spectral analysis (see later, Chapter 4).

The emphasis in this early work on frequency analysis was directed towards finding a plausible model for the phenomenon under investigation. Slutsky [4] suggested a moving-average and regressive series models, both of which appeared to represent the behaviour of many physical variables. These particular models have reappeared recently in connection with digital filters but are otherwise little used as spectral representations. Both of Slutsky's series either individually or in combination played a useful part in spectral representation until the development of the more generalised theories of Wiener [5] and others which reduce the dependence on specific models of the time series. Wiener considered the generation of a series of harmonically related terms

$$x(t) = \sum_{i=1}^{\infty} A_i \exp(i,\omega_i,t) \tag{1.14}$$

which is essentially a Fourier-type representation and appears in one form or another in nearly all the methods of analysis later proposed for the frequency domain. In his paper on generalised harmonic analysis Wiener laid a solid mathematical foundation for the analysis of random processes which has had far-reaching consequences in present-day work on communication, control and information theory.

The origins of the present methods of spectral analysis, based on a Fourier representation of a series, are due mainly to the work of Tukey and Bartlett [6, 7]. They realised that a relationship must exist between the easily calculable covariance function of the time-history and its power spectrum.

In fact, the variance is simply

$$V(x) = \overline{[x(t)-\overline{x}] [x(t+\tau)-\overline{x}(t+\tau)]} \tag{1.15}$$

and the power spectrum is the cosine Fourier transform of this, i.e.

$$S(\omega) = \frac{1}{2\pi} \int_{-\infty}^{\infty} V(x) \cos \omega\tau . d\tau \tag{1.16}$$

The relationship had been rigorously established earlier by Wiener, but lacked practical application to the problem of power spectral density evaluation.

In its later form the zero-mean auto-correlation replaced the auto-covariance and the method of calculating power spectrum from the cosine transform of the mean-lagged products proved the cornerstone of spectral signal processing for a number of years. One reason for the success of this method was the reduction in calculation time. Despite the two-stage method of calculation via the auto-correlation, it proved quicker to do this rather than to calculate all of the Fourier coefficients directly.

A certain measure of success has been obtained with other more direct methods. Of particular interest is the technique of **complex demodulation** [8] in which the time-history series, $x(t)$ is multiplied by $\sin \omega_0 t$ and $\cos \omega_0 t$, giving

$$x(t) \sin \omega_0 t, \qquad x(t) \cos \omega_0 t \tag{1.17}$$

Applying the resultant values through a low-pass filter gives the two phase-related terms

$$a = F(x_t . \sin \omega_0 t), \qquad b = F(x_t . \cos \omega_0 t) \tag{1.18}$$

The complex spectrum can be obtained as an amplitude function

$$A = 2(a^2 + b^2)^{1/2} \tag{1.19}$$

and a phase angle

$$\theta = \tan^{-1}(a/b) \qquad (1.20)$$

for a given angular frequency, ω and time, t.

Both the calculation via the mean-lagged product and that of complex demodulation were carried out almost exclusively using analog methods. Integration is both fast and cheap using the operational amplifier and digital computation proved an expensive way to the frequency spectrum.

A considerable change in the methods of calculation for the frequency domain was brought about, however, by the rediscovery of the fast Fourier transform algorithm in the 1960s [9]. Calculation reverted to the direct method and the digital computer became used extensively for this purpose. To calculate all the Fourier components for N sample values, using the algorithm, requires approximately $N.\log_2 N$ complex multiply-add operations compared with N^2 operations using the previous direct transform methods. For a long series this results in a reduction by several hundreds or even thousands of times in computer run-time.

The speed of the fast Fourier transform enabled it to be used not only for the direct calculation of the power spectral density, but also, via the Wiener theorem in the calculation of auto-correlation. (This, it will be noted, is a reversal in the order of two-stage calculations brought about solely by the speed of the fast Fourier transform algorithm.) Its advantages in convolution and in digital filtering were quickly realised and described in several early papers, notably those of Stockham and Helms [10,11].

Whilst spectral methods have been most widely applied and understood, particularly in the engineering field, more recent developments have emphasised the value of correlation and convolution in the time domain. This is due to the value of these functions in the analysis of the behaviour of physical systems and the use of auto- and cross-correlation for signal identification purposes. Correlation is used particularly as a means of improving the signal-to-noise ratio enabling the extraction of periodic signals immersed in a noisy signal, such as would be obtained, for example, from the reception of long-range radar echoes. The technique used is to consider the sum of the cross-products of the signal, x(t) and a delayed replica of itself, $x(t-\tau)$ over a finite time interval. This summation is repeated over a range of delay values with the result that the cyclic similarity between the two signals for the periodic components is preserved whilst the summation tends to zero for the random noise element. The use of cross-correlation to determine the impulse response of a system is a technique for which the applications appear to be growing. It is now fairly commonplace to inject a perturbation into a system (mechanical or otherwise), and to determine the system impulse response or transfer function by cross-correlating the system output changes with the input perturbation (noise). The value of such a method of measurement lies in the fact that little disturbance need be made to the system operating normally.

A somewhat similar use of cross-correlation is its use in establishing vibration or shock transmission routes through a structure. From a knowledge of the initial vibration frequency and amplitude at entry point the modification to this signal during its passage can be determined at a suitable exit point and thus provide information regarding the characteristics of the transmission route.

Correlation and its derivatives have also been used extensively in the analysis and study of vocal mechanisms. A particular form of spectrum having close similarities with auto-correlation, known as the **cepstrum**, is valuable for this purpose [12]. Whereas auto-correlation may be defined as the Fourier transform of the power spectrum the cepstrum is defined as the transform of the logarithm of the power spectrum. Since the log-power spectrum is an even function this definition is equivalent to the square of the cosine transform of the log-power spectrum, viz.

$$K_x(\tau) = \left| \int_0^\infty \log_e \left| X(\omega) \right|^2 \cos (\omega\tau) . d\omega \right|^2 \qquad (1.21)$$

This definition permits the unambiguous indication of the rate of repetition for the component parts of periodic signals which are very non-linear in form. Such signals are rich in harmonically related terms and an auto-correlogram would present a complicated indication in the time domain due to the profusion of extracted periodicities.

Statistical methods of analysis involving the modelling of time series have been developed recently as alternatives to transform analysis. These are known as the **maximum entropy** and **Pisarenko** methods of spectral decomposition [13, 14]. In both methods the uniform characteristics of white noise form the basis for the signal being modelled as a spectral estimate of the original signal from whose auto-correlation coefficients the modelling parameters are taken. The maximum entropy method applies an auto-regressive process to these auto-correlation coefficients. The resulting series provides a modifying function to the basic white noise to produce a model of the original signal. The motivation for this apparently complicated procedure is to improve the effective signal-to-noise ratio of the original signal. Unfortunately the process tends to reproduce only the most prominent spectral characteristics of the signal so that its application is limited to certain special signals where this performance is found acceptable. The Pisarenko method is fundamentally similar but enables a much finer spectral resolution to be achieved. Both methods are discussed in Chapter 4.

The analysis methods referred to above all assume a stationary and often an ergodic signal, so that it is sufficient to consider the signal as a function of time or frequency alone. In many physical situations both must be considered and various definitions of short-term or instantaneous spectra for time-varying signals have been proposed.

An early view was to consider the energy contained in the signal by way of a double integral form of the Wiener theorem

$$S_x(\omega_1,\omega_2) = \int\limits_{-\infty}^{\infty} \int\limits_{-\infty}^{\infty} R_x(t_1,t_2) \exp [j(\omega_1 t_1 - \omega_2 t_2)] \, dt_1 . dt_2 \qquad (1.22)$$

which has been termed the **generalised spectral density function**. A difficulty lies in the interpretation of this result due to the lack of a physical meaning relating it to a stationary form of analysis.

Other forms of analysis, notably those of Priestley [15] are applicable to certain single realisations of a process, and are closer to the single-point idea of spectral density. This estimate is referred to as **evolutionary spectral analysis** and expressed in terms of energy and frequency density by the function

$$f(t,\omega) = |A(t,\omega)|^2 f(\omega) \qquad (1.23)$$

The function of time and frequency $A(t,\omega)$ performs the operation of filtering the signal over a limited time and frequency bandwidth so that the signal may be considered stationary over this period. It is therefore confined to functions that are found to change slowly with time.

The optimum width of the time and frequency windows implied in equation (1.23) is difficult to obtain analytically and the method can result in a large amount of regression analysis being required, using the digital computer to arrive at optimum window values.

Amplitude domain analysis in the form of probability estimates has proved useful in predicting the behaviour of systems to environmental stress. For practical reasons an equivalence between sinusoidal and random inputs for vibration testing is often taken as valid, implying that the process under test behaves in a Gaussian way. This may not always be the case so that initial probability density measurements now form an essential element in the evaluation of physical systems subject to shock and vibration stress analysis.

Other amplitude domain measurements are statistical peak counts and zero-crossing measurements which are particularly relevant in assessing susceptibility to fatigue damage or in establishing boundary conditions.

1.4   Basic Properties of the Signal

Assuming that the signal is stationary and ergodic then its properties can be defined in probabilistic terms for the amplitude, time or frequency domain.

Brief descriptions of these properties will be given here to serve as an introduction to later chapters where they will be considered further and methods of deriving them described in some detail.

1.4.1 AMPLITUDE DOMAIN ESTIMATES

The simplest form of description for a random signal is given by its **mean-square value**, defined as

$$\bar{x}^2 = \lim_{T \to \infty} \frac{1}{T} \int_0^T x^2(t) . dt \qquad (1.24)$$

or the positive square root of this quantity known as the **root-mean-square value**.

These values will give an indication of the amplitude effect of the signal but will not give sufficient information for the detailed understanding of the variable nature of the process. To do this the behaviour of the signal in terms of the probability of its amplitude value exceeding a given level, or lying between specified levels, must be determined. Referring to Fig. 1.8,

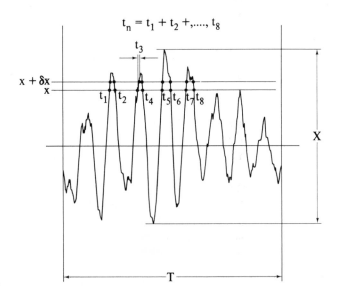

Fig. 1.8   Determination of probability

the fraction t_n of the total record time, T, that the signal lies between discrete levels x and x + δx over the complete dynamic amplitude excursion X of the signal is given as

$$t_n = \sum_{p=0}^{p=k} t_p (x, x + \delta x, n). \qquad (1.25)$$

$$p = 0, 1, \ldots, k$$

where n indicates the number of levels into which the signal amplitude is divided, and t_p indicates the dwell period for each excursion the signal makes between pairs of levels.

The **probability density function** describes the probability that the signal will be found within a given range, x and x + δx and is normalised by δx to give a density function

$$p(x) = \lim_{T \to \infty} \frac{t_n}{T \delta x} \qquad (1.26)$$

The ratio given in equation (1.26) will approach an exact probability value as the record length extends to infinity. If the number of possible discrete values for x is great, then a probability function with a large domain is required to describe the function x(t). In such cases it is often more convenient to use the **probability distribution function**, which describes the probability that the variable will assume a value less than or greater than x. This is given by the integral of the probability density function as

$$P(x) = \int_{-\infty}^{x} p(x) \cdot dx \qquad (1.27)$$

Examples of these functions are given in Fig. 1.9.

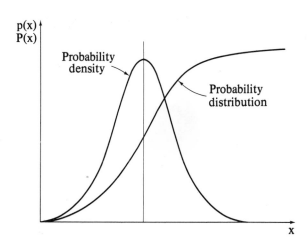

Fig. 1.9   Probability functions

These definitions can be extended to include the probability of two associated sets of data being in a given condition at the same time. The **joint probability density function** defines the probability that x(t) will be found within the range x and x+δx whilst y(t) simultaneously assumes a value within the range y and y+δy.

The time t_n now defines the fraction of the total time that x(t) and y(t) simultaneously fall within these ranges over the record length T, so that

$$p(x,y) = \lim_{T \to \infty} \frac{t_n}{T, \delta x, \delta y} \qquad (1.28)$$

and the **joint probability distribution function** is

$$P(x,y) = \int_{-\infty}^{x} \int_{-\infty}^{y} p(x,y)dx,dy \qquad (1.29)$$

1.4.2 TIME DOMAIN ESTIMATES

Simple averages and probability measurements fail to tell us anything about the periodic behaviour of the signal. Statistically this information can be obtained by making measurements of the amplitude of the signal at two times, separated by a delay τ, finding their product, and averaging over the time of the record (see Fig. 1.10). This procedure is known as **auto-correlation** and results in the **auto-correlation function**

$$R_x(\tau) = \lim_{T \to \infty} \frac{1}{T} \int_{0}^{T} x(t).x(t+\tau).d\tau \qquad (1.30)$$

It is expressed in graphical form as an **auto-correllogram** for $R_x(\tau)$ against time or delay (τ).

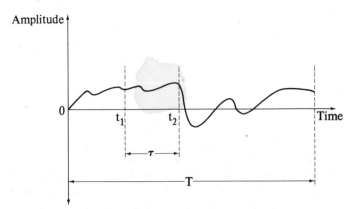

Fig. 1.10   Auto-correlation

The main value of the auto-correllogram is its ability to reveal the presence of periodicity in a random signal. For a completely random signal even a slight variation in delay will result in a reduction of the product, $x(t)x(t+\tau)$ to a very small value so that a large value (normalised to unity), will only be obtained at or close to $\tau=0$. This will not be the case for periodic components where a large shift in τ equal to half a period will be required before any substantial change to $R(\tau)$ will be apparent. Examples shown in Fig. 1.11 indicate clearly the value of the auto-correllogram in differentiating between random and periodic signals.

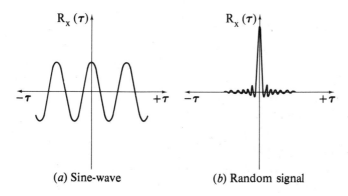

(*a*) Sine-wave   (*b*) Random signal

Fig. 1.11   Examples of auto-correlation

A similar result is obtained if the measurements are taken from two signals, $x(t)$ and $y(t)$ at times separated by τ. The relationship is expressed as the **cross-correlation function**

$$R_{xy}(\tau) = \lim_{T\to\infty} \frac{1}{T} \int_0^T x(t) . y(t+\tau) . dt \qquad (1.31)$$

This gives an indication of the joint properties which the two signals may share.

1.4.3 FREQUENCY DOMAIN ESTIMATES

Although information on the spectral decomposition of a signal can be derived from the auto-correllogram a more useful description is obtained by a direct plot of its spectral content. Fourier series analysis can be applied to periodic functions and values of the Fourier coefficients will give directly the peak amplitude of the related harmonics contained within the signal. This form of analysis, however, is inadmissible for random signals which do not necessarily have harmonically related components. Instead a measure of the relative amplitude of the frequency components is obtained via the Fourier transform as

$$X(f) = \int_{-\infty}^{\infty} x(t)\exp(-j\omega t).dt \quad (\omega = 2\pi f) \tag{1.32}$$

The mean-square value is taken to obtain a measure of the instantaneous power at a given frequency over the record length T, viz.

$$S(f) = \lim_{T \to \infty} \frac{1}{T} \int_{0}^{T} \frac{x_B(t)^2}{B}.dt \tag{1.33}$$

where $x_B(t)^2$ is the instantaneous square of the signal contained within a bandwidth, B. This is related to the square of the modulus of the Fourier transform, $|X(f)|^2$ for a given number of degrees of freedom appropriate for the effective filter bandwidth, B (see Chapter 4). Equation (1.33) gives the **power spectral density** of the signal obtained by averaging the squared values of the signal over a narrow bandwidth, B. An example of such a spectrum is given in Fig. 1.12 for a random signal containing several strong periodic components.

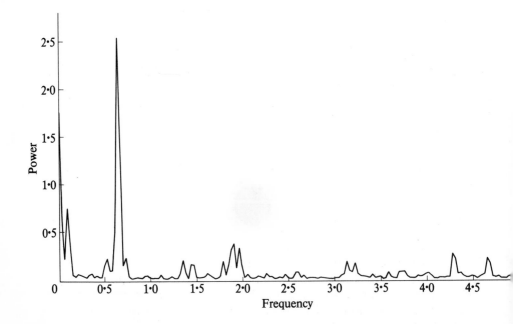

Fig. 1.12   Power spectral density

The relationship mentioned earlier, between the power spectral density function and the auto-correlation function, is an important one in terms of practical measurement methods. The two functions can be shown to be Fourier transforms of each other, viz.

$$S(f) = \int_{-\infty}^{\infty} R(\tau) \exp(-j\omega\tau).d\tau \tag{1.34}$$

and

$$R(\tau) = \int_{-\infty}^{\infty} S(f) \exp(j\omega\tau).df \tag{1.35}$$

In a practical case the relationships for real values at positive frequencies only are given as

$$S(f) = \int_{0}^{\infty} R(\tau) \cos \omega\tau.d\tau \tag{1.36}$$

$$R(\tau) = \int_{0}^{\infty} S(f) \cos \omega\tau.df \tag{1.37}$$

Equations (1.34 to (1.37) are known as **Wiener-Khintchine relationships**.

Similar relationships can be derived for the cross-power spectrum for two signals derived from the same physical system. However, since the cross-correlation function is not an even function, its physically realisable one-sided cross-spectral density function, $S_{xy}(f)$ is defined by

$$S_{xy}(f) = C_{xy}(f) - j Q_{xy}(f) \tag{1.38}$$

where $C_{xy}(f)$ is the co-spectral density and
$Q_{xy}(f)$ is the quadrature spectral density expressed as

$$C_{xy}(f) = \int_{0}^{\infty} [R_{xy}(\tau)+R_{xy}(-\tau)] \cos(\omega\tau)d\tau \tag{1.39}$$

and

$$Q_{xy}(f) = \int_{0}^{\infty} [R_{xy}(\tau)+R_{xy}(-\tau)] \sin(\omega\tau)d\tau \tag{1.40}$$

The cross-power spectrum can be considered as a plot of the power contained in the product of the two time-histories, $x(t)$ and $y(t)$, taken over successive narrow-band frequency intervals.

1.5 Procedure for Analysis

The dependence of the method of analysis on the environmental conditions under which the signal becomes available has been mentioned earlier. This applies particularly to the conversion of the physical quantity into an electrical signal. Any set of procedures for analysis of transducer-derived time-histories must include pre-processing and calibration of the signal in order to take this into account.

Pre-processing plays a major role in conditioning the signal into a form suitable for analysis. For example the presence of a large amplitude slowly-varying trend can prevent effective analysis of small rapid changes in the signal by restricting the usable dynamic range. Trend removal, which in this case amounts to simple filtering, can improve considerably this situation. Filtering is also required prior to digitisation or decimation of a discrete time series. Calibration implies known information concerning acquisition of the data. All of this information, with the pre-processing conditions and physical characteristics of the system under study, is required to permit interpretation of the signal. The importance of established procedures, and particularly documentation procedures, cannot be stressed too highly, since a time-history alone is valueless unless associated with the necessary information regarding its manner of acquisition.

Digital computer methods are highly repeatable and extremely accurate as regards the calculations involved. Cost is no longer a major limitation with fast Fourier transform algorithms and efficient hardware. A more serious difficulty is that of the initial conversion of the continuous signal into a sampled form. Digitisation problems are discussed in Section 1.5.2.

Tests for signal characteristics are also required as part of the analysis procedure. It is important to establish a testing procedure to detect the presence of periodic components in an otherwise random signal. The most powerful method is the use of the auto-correlation function. Measurements taken from the auto-correlogram can be used to define a region for investigation using power spectral density analysis to determine the value of the detected periodic components. A search for these will often reveal sinusoidal peaks by their relationship to the bandwidth of the analysis filter.

If the periodic component is truly sinusoidal its transformed value will be a delta function. Consequently if we consider spectral analysis through the use of a band-pass filter then varying the width of the filter will not affect the instantaneous power, $X_B(t)^2$, although the peak value of $S_x(\omega)$ will vary inversely with the filter bandwidth, B (equation 1.33) thus giving an indication of sinusoidal content.

It will also be necessary to reject unwanted data from the signal before analysis commences. Data acquisition methods will almost always result in extraneous information being included, either in the form of an extended record including pre- and post-transient recording, or in extended bandwidth. Time-compression or decimation techniques can result in reduction in the quantity of data to be stored and processed and will be justified in many cases.

1.5.1   REQUIRED LENGTH OF RECORD

An important factor determining the accuracy available for the statistical estimates is the amount of data available. Each method of analysis will define its own minimum sample length for a given estimation accuracy, and these lengths will be different.

It will be shown later that for all the basic properties of the signal discussed in Section 1.4 the record length required, T, is inversely proportional to twice the analysis bandwidth, B, multiplied by the square of the estimation error ϵ, and will also be proportional to a constant K, whose value is dependent on the property being measured [16], i.e.

$$T = \frac{K}{2B\epsilon^2} \tag{1.41}$$

For mean-square value and power spectral density estimates a value of K = 2 can be assumed. Half this record length is required for correlation estimates where measurements take place around the peak value. As the lag (τ) increases then the record length required for equal error increases, i.e.

$$K = 1 + \left(\frac{R_x(0)}{R_x(\tau)}\right)^2 \tag{1.42}$$

Record length requirement for mean-value estimation is dependent on the square of the ratios of the standard deviation σ, to the mean value, i.e.

$$K = \left(\frac{\sigma}{\bar{x}}\right)^2 \tag{1.43}$$

and therefore depends on the dynamic range of the signal.

Probability estimates require the longest sample of the record since we are equally interested in the probability of very small and very large values and may need a long record to collect sufficient high amplitude values to form a statistically viable estimate. For probability estimates

$$K = \frac{1}{A \cdot p(x)} \tag{1.44}$$

where A is the amplitude resolution width, δx, over the resolution and p(x) is the probability density estimate.

1.5.2 DIGITISATION OF CONTINUOUS INFORMATION

Conversion of a continuous analog signal to a discrete form suitable for input and processing in a digital computer involves sampling in the time domain, quantisation in the amplitude domain, and coding the resulting information into digital form.

These three processes all impose limitations on the data obtained and can give rise to various sources of errors which will be considered in this section.

Sampling of the analog signal involves the selection of a series of narrow impulses or 'slices' of the signal, spaced at equal time intervals. A unique number is ascribed to each impulse and represents the mean amplitude of the sample taken over the area of the impulse. Sampling in the time domain is assumed here to take place at a uniform rate. This need not be so and there may be certain advantages in sampling in accordance with a linear function of time (e.g. a sinusoidal function known as **cyclic rate sampling** [17]), or at a rate related to the signal being sampled, or even in a random manner. These methods are not of direct value for signal processing purposes although there are advantages in sampling in a random manner to overcome aliasing effects and aliasing noise [18]. Only the **uniform** method of sampling will be considered in this chapter.

Ideally, we would like each sample to be taken over an infinitely short period of time but, in a practical case, it is necessary to estimate an averaged quantity over the sampling period. The length of time over which the data are averaged is known as the **aperture**. The width of this is important and to obtain a minimum conversion error the aperture width should be small compared with the sampling period. Aperture errors are minimised by the use of very fast multiplexors and analog-to-digital converter devices so that the time shift in the input−output sequences is negligible compared with the smallest signal period. We have assumed that sampling of the continuous signal is to be carried out at regular intervals. Unless this is done spurious frequencies related to the uncertainty in time location of the sampling interval will be introduced into the sampled data. This uncertainty is referred to as jitter of the sampling pulse and when present can introduce progressive errors in the phase information associated with high frequency signals.

1.5.3 ALIASING

The most important of the difficulties that arise when a continuous signal is sampled is undoubtedly that of **aliasing**. The nature of this problem is illustrated in Fig. 1.13 which shows that the same set of sampled data points can describe a number of time-series histories which are indistinguishable to the digital computer. Any time-history can be expressed as the summation of a series of cosinusoids. This enables a simplified model of a real signal to be assumed, and provide a means of quantitatively analysing this aliasing effect. Consider the sampled time-history shown in Fig. 1.14. Here the signal waveform, x(t), is sampled at regular intervals, h, and the amplitude of the points plotted as before. We will assume the function x(t) to represent a cosinusoid of frequency f_0. The same points could equally well be taken to represent a cosinusoid of frequency f_1 or f_2 which are multiples of f_0. These 'aliased' frequencies are obviously related to the sampling period h and the aliased sequence of frequencies can be written as

$$f_0, \quad \frac{1}{h}-f_0 \quad , \quad \frac{1}{h}+f_0 \quad , \quad \frac{2}{h}-f_0 \quad , \quad \frac{2}{h}+f_0 \quad ,\dots, \quad (1.45)$$

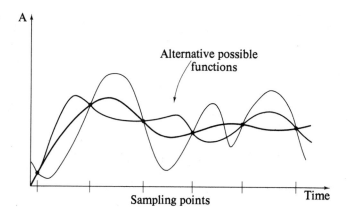

Fig. 1.13 Aliasing

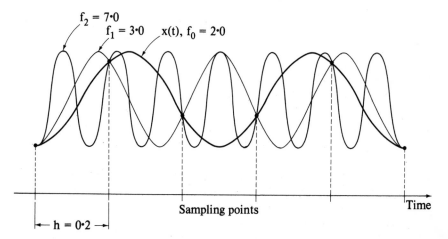

Fig. 1.14 Analysis of the aliasing effect

The fundamental frequency, f_0, is known as the **principal alias**. The range of frequencies below which this effect is not present extends from $f_0 = 0$ to $f_0 = f_N$. This maximum frequency, f_N, is known as the folding or **Nyquist frequency** and is also referred to as defining a frequency limit—the Shannon limit [19] to the sampled data, above which an unambiguous reconstruction of the signal is not possible. It can be described as the lowest frequency coinciding with one of its own aliases. This can be seen from Fig. 1.14 and is clearly $(1/h = f_0)$. Thus, given a signal containing no frequency components at and beyond a frequency f_N, which we may call the bandwidth, B, of the signal, then the lowest sampling frequency necessary to preserve the information contained in a sampled version of this signal is given as $f_S \geqslant 2B$ or since $f_S = 1/h$, then $B = \frac{1}{2h}$. This is known as the **sampling theorem**. It follows that for a given frequency spectrum, the individual frequency components lying between $f = 0$ and $f = B$ can be separately examined, but if the signal contains components having frequencies $f > B$ they will not be distinguishable. Since in a practical case, these higher order components can contribute some power to the frequency spectrum, their contribution will be indistinguishable from those lying between $f = 0$ and $f = B$.

A further consequence of sampling a continuous signal is that its spectrum is now represented by a set of identical coefficients, repeated for successive frequency bands equal in width to the value of the Nyquist frequency. The repetition must be considered when manipulating the generated spectral series as we shall see later when digital filters are discussed.

The sampling theorem also applies to the time domain so that if the signal of interest lies within a frequency band extending from 0 to B Hz then the minimum length of record necessary in order that we may recover this signal from the sampled data is

$$T \geqslant \frac{1}{2B} \text{ sec} \qquad (1.46)$$

which is known as **Rayleigh's criterion**.

From the preceding discussion we see that the sampling theorem consists of two parts which may be restated as:

1. Signals having a finite bandwidth up to and including B Hz can be completely described by specifying the values of the time history series at particular instants of time separated by $\frac{1}{2B}$ sec.
2. If the signal is band-limited and contains no frequency greater than B Hz, it is theoretically possible to recover completely the original signal from a sampled version when the sampling interval is equal to or smaller than $\frac{1}{2B}$ sec.

This concept of finite bandwidth is important. Consider a signal x(t) to contain no frequencies higher than B Hz. We can represent this in the frequency domain by its Fourier series, X(f) (defined in Chapter 3).

$$X(f) = \frac{1}{2B} \sum_{n=-\infty}^{+\infty} C_n \exp\left(\frac{-j\,\pi nf}{B}\right)$$

$$-B < f \leqslant B$$

where

$$C_n = \frac{1}{2B} \int_{-B}^{+B} X(f).\exp\left(\frac{j\,\pi nf}{B}\right).df \tag{1.47}$$

and represents the complex Fourier amplitudes of the series. Now X(f) fully defines the spectrum so that the equivalent time function x(t) can be obtained from the inverse transform

$$x(t) = \int_{-B}^{+B} X(f).\exp(j\omega t).df \tag{1.48}$$

If t is defined as $\frac{1}{2B}$ then equation (1.48) becomes

$$x\left\{\frac{n}{2B}\right\} h = \int_{-B}^{+B} X(f).\exp\left(\frac{j\,\pi nf}{B}\right).df = 2BC_n \tag{1.49}$$

From equations (1.47) and (1.49) we can deduce that if a band-limited function X(f) is sampled at times $t = \frac{nB}{2}$ the original signal is completely recoverable from the sampled signal with no loss of information. Hence the sampling period must be $h \leqslant \frac{1}{2B}$.

In practice this recovery is not perfect and the reason lies in the inadequacy of filter design. This may be seen if we consider the Fourier representation of a spectrum of limited bandwidth (Fig. 1.15) shown here as a two-sided spectrum. This necessarily includes an infinite series of spectra on

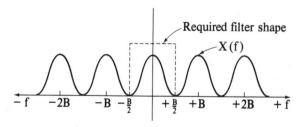

Fig. 1.15   Fourier representation of a limited bandwidth signal

either side of the original spectrum so that a filter having the characteristics, shown dotted in Fig. 1.15, is required for full recovery of the original spectrum. Such perfect filter characteristics are unattainable and some distortion due to sampling is therefore inevitable.

From the considerations given above, it is seen to be essential that the signal be subjected to a low-pass filtering operation prior to digitisation to ensure that all frequencies greater than $B = f_N$ are excluded. This is necessary, not only to avoid the aliasing effects of the actual signal content, but also to reduce the contribution of higher frequency order noise components to the digitised data, which will otherwise be accepted as noise components falling within the Nyquist bandwidth. The inadequacy of practical filters used for band-limiting modifies our choice of sampling rate such that a rather higher rate is required than is suggested by the sampling theorem. We may, for example, find that the necessary attenuation (equal to the required dynamic range of the measurement) will not be reached until a frequency two octaves higher than the cut-off frequency. Commonly the sampling rate is set at $1 \cdot 25$ times the filter cut-off frequency. This is because the addition or 'folding' into the accepted band is actually vector addition. Where the phase characteristics of the higher frequency signal are precisely known then a lower choice of sampling frequency can be tolerated.

1.5.4 QUANTISATION

Representation of a variable amplitude series of discrete sample values as a limited series of discrete numbers is termed quantisation. The process can only be an approximation since, whilst the original signal can assume an infinite number of states, the number of bits in a digital representation is limited. The numerical values of the quantised variable is represented by some form of binary code to permit entry into a digital computer or device.

In broad terms, quantisation is a non-linear operation that is carried out whenever a physical quantity is represented numerically. The resultant numerical value is given as an integer corresponding to the nearest whole number of units.

This is expressed by the transfer characteristics of a quantiser shown in Fig. 1.16. An input value lying between the midpoint values of two consecutive unit values will produce an output value at the level corresponding to the higher of the two values.

Applying this transfer characteristic to a continuous function, given in Fig. 1.17, shows that this function may be represented in quantised form by a series of numerical values having integer values. This process suggests that quantisation is like sampling in amplitude, and, in fact, Widrow [20] has referred to quantisation as a sampling process that acts not upon the function itself, but upon its probability density distribution.

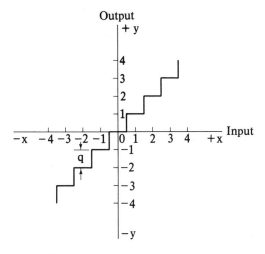

Fig. 1.16   Quantiser transfer characteristic

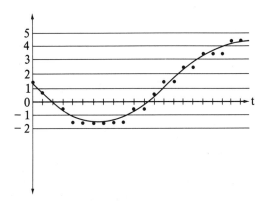

Fig. 1.17   Sampling and quantisation

A sampling theorem of quantisation also exists which states that if the dynamic range of the input signal extends over several intervals then the statistical properties of the signal are recoverable. This way of looking at the process of quantisation in statistical terms is useful in that it enables a value to be placed on the error of the process and will be referred to later.

Both sampling and quantisation introduce distortions into the treated continuous signal. The distortion or error introduced into the process of quantising an already time-sampled signal gives rise to a form of additional noise which is considered below.

The continuous signal x(t) is sampled in time and may be regarded as the process of modulating x(t) by a series of impulses spaced uniformly along a time-axis, $D(n) = \Sigma(\delta t - nh)$. This gives a convolution summation

$$x_q(t) = \sum_{n=-\infty}^{\infty} x(t)\delta(t-nh) \qquad (1.50)$$

where $(t-nh)$ represents an impulse of unit area (**delta function**) delayed by nh seconds and h is the sampling interval. As we shall derive later, the spectrum for a rectangular pulse series can be obtained as

$$D(n) = AB/h\left\{ \frac{\sin n\pi(B/h)}{n\pi(B/h)} \right\}$$

where A represents the pulse height and B its width. For a delta function $B \to 0$ when $AB = 1$ so that

$$\lim_{B \to 0} D(n) = \frac{1}{h}$$

and the inverse Fourier transform of D(n) can be substituted in equation (1.50) to represent the delta series in the time domain, i.e.

$$x_q(t) = x(t)\ \frac{1}{h} \sum_{h=-\infty}^{\infty} \exp(j2\pi nt/h) \qquad (1.51)$$

Thus the quantised signal can be expressed as the product of the Fourier series for the impulse train and the continuous signal, x(t). It will consist of an infinite number of harmonically related sinusoidal carriers, having a uniform frequency spacing 1/h, which are modulated by the input signal to produce a pattern of identical sidebands about each carrier frequency. It is only necessary to filter the modulated signal with a low-pass filter of cut-off frequency close to that of the modulating signal to recover the original signal. This assumes, of course, that the signal is band-limited so as to exclude frequencies higher than the Nyquist limit, 1/2h, as discussed previously. Thus the frequency characteristics of the quantised signal enable a mechanism for recovering the original continuous signal to be realised.

The effect of quantisation on the original signal is, as we said earlier, to introduce an added noise to the signal. This is illustrated in Fig. 1.18 which considers the signal to be applied via a non-phase-shifting device having unit gain into which is introduced a noise input. It can be shown [20] that this noise has the characteristics of white noise which has a rectangular probability distribution.

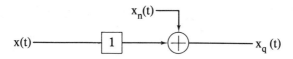

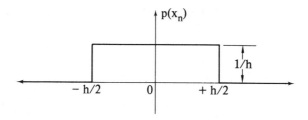

Fig. 1.18   Equivalent representation of the quantisation process

Thus we can represent the quantised output as the sum of the input signal, x(t) and noise component n(t)

$$y_q(t) = x(t) + n(t) \tag{1.52}$$

where n(t) can be precisely defined if the dynamic range of the input and number of quantising levels are known.

This noise represents a quantisation error, shown in Fig. 1.19 as a function of the quantiser input. The quantisation error is clearly related to the least value of quantisation q, and can be given as the ratio of the magnitude of the signal at the point of sampling expressed as a sample number A, and the approximate number resulting from quantisation given as a power of 2, viz.

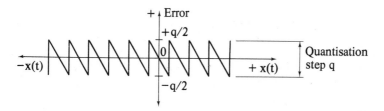

Fig. 1.19   Quantisation noise characteristic

$$\epsilon = \pm \frac{A}{2} \cdot \frac{1}{2^N} = \pm \frac{q}{2} \tag{1.53}$$

To assess the value of this error let $p(x)$ be the quantisation error probability density function defined by

$$p(x) = 1 \text{ for } -0 \cdot 50 \geqslant x \geqslant + 0 \cdot 50$$
$$= 0 \text{ otherwise.}$$

The variance for this error is

$$\sigma^2_x = q^2 \int_{-\infty}^{\infty} (x - \bar{x})^2 \, p(x)dx$$

and since the mean value must be zero

$$\sigma^2 = q^2 \int_{-0 \cdot 5}^{0 \cdot 5} x^2 dx = q^2 \left(\frac{x^3}{3} \right)_{-0 \cdot 5}^{0 \cdot 5} = \frac{q^2}{12} \qquad (1.54)$$

giving a standard deviation for a single unit of quantisation of $\sigma_x = 1/\sqrt{12} = 0 \cdot 29$ level units. This gives a value for the quantisation noise which is added to the desired signal. As an example if we quantise a signal to 256 level units (i.e. 2^8 or 8 binary bits), then the signal-to-noise ratio at the output of the quantiser will be

$$\frac{256}{0 \cdot 29} = 1000 \text{ or } 60 \text{ decibels}$$

Most A/D converters now available have a dynamic range of 16 bits so that this order of accuracy is easily realisable. To express this error in more general form, we have, from equations (1.53) and (1.54):

$$\sigma^2_x = \left(\frac{A}{12} \right) 2^{-N} \qquad (1.55)$$

or

$$A^2 = 12 \, \sigma^2_x \, 2^N, \qquad (1.56)$$

but the dynamic range of the input signal expressed in decibels is $R = 20 \log_{10} A$ db so that from equation (1.56)

$$R = 10 \log_{10} 12 + 20 \log_{10} \sigma_x + 10 \, N \log_{10} 2$$

and

$$N \geqslant \frac{1}{3} [R - 10(1 + 2 \log_{10} \sigma_x)] \qquad (1.57)$$

This allows the number of binary bits, N, or word length of the input signal to be defined for a given dynamic range (in decibels). The quantisation error must be expressed as a standard deviation, σ_x.

1.6 Signal Processing Mathematics*

Certain concepts in the field of mathematics are essential for a clear understanding of signal processing techniques. This section can do little more than introduce the most relevant of these and provide the necessary references for further reading. Here we will develop the theme of linear mathematics, considering the function of eigenvectors, and matrices, which play an important part in signal theory. The treatment is not rigorous but should provide a sufficiently adequate background for the material that follows later in this book.

1.6.1 LINEAR MATHEMATICS

We start with a simple case. The relation

$$y = Ax + B \qquad (1.58)$$

is linear, because if we plot y against x the figure is a straight line.

It should be noted that A and B may change from example to example. They are adjustable parameters. However, they do not change with x. (Otherwise the relation is no longer linear.) In this sense they are 'constant'.

Equation (1.58) expresses a single value y in terms of a single value x. They are what we call *scalars*. Quite often, we come into contact with **vectors** and **functions**, which consist of many different values. Further, each value of y may vary with *every* value of x. First consider the case when y is scalar but x is a vector. Then

$$y = \sum_i A_i x_i \qquad (1.59)$$

is a linear relation. We can convert equation (1.59) into the form of equation (1.58) for any particular component of x, say x_p as

$$y = A_p x_p + B \text{ with } B = \sum_{i \neq p} A_i x_i$$

Thus, y is a linear function of each value contained in the vector x. When x is a function of t, then integration replaces addition, and, we have, again for a scalar y

$$y = \int A(t) x(t) dt \qquad (1.60)$$

Now consider the case when y also contains multiple values, and each value of y is linearly related to each value of x. Different values of y may of course have different coefficients. Thus we have

$$y_k = \sum_i A_{ik} x_i \tag{1.61}$$

i.e., each term y_k is a linear function of each x_i with coefficient A_{ik}. Similarly we have for functions of t

$$y(t) = \int A(t, t') x(t') dt' \tag{1.62}$$

The set of coefficients in equation (1.61) form a **matrix**, which is just a common name for a two-dimensional table of numbers. The set of coefficients in equation (1.62) form a **kernel**. Both are **linear operators**, because they convert a vector or a function x into another y by means of linear arithmetic. Frequently, we have physical systems which respond to external stimulus in an approximately linear way, that is, the response to the sum of several stimuli is just the sum of the individual stimuli. These are called linear systems. Up to a point, almost any physical system is approximately linear. Note that equation (1.62) satisfies the criterion of linear systems: doubling x doubles y, and the 'response' to 'stimulus' $[x_1(t)x_2(t)]$ is just

$$\int A(t,t') x_1(t') dt' + \int A(t,t') x_2(t') dt'$$

the sum of the individual responses. In fact, by choosing $A(t,t')$ appropriately we can approximate mathematically the input–output relation of most linear systems.

A particularly important kind of linear system is the linear, time-invariant system, for which $A(t,t')$ obeys the restriction

$$A(t,t') = A(t - t') \tag{1.63}$$

Referring to Fig. 1.20, let us first suppose that the stimulus is a Dirac impulse, $\delta(t)$ at time 0, where $\delta(t) = 0$ except at $t = 0$. The response, $A(t)$, starts at $t = 0$, builds up to some maximum s. Now suppose we apply an impulse at time $t = s$. This is described by $\delta(t - s)$, since $\delta(t - s) = 0$ except at $t = s$. The response is A delayed by time s. It is thus $A(t - s)$. Now if both impulses are applied, the response is the sum of the individual responses since the system is considered to be linear. Finally, an arbitrary input x(t) can be expressed as a continuous train of impulses at different times. In equation form

$$x(t) = \int_0^t \delta(t - s) x(s) ds \tag{1.64}$$

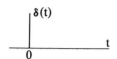

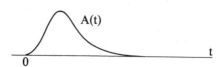

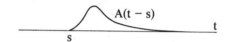

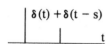
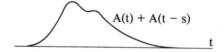

Fig. 1.20   System response to a Dirac impulse

The response is in fact the combination of the individual responses to $x(s)\delta(t - s)$, or

$$y(t) = \int_0^t A(t - s)x(s)ds \qquad (1.65)$$

which expresses equation (1.62) after applying the restriction equation (1.63). We will see later (Chapter 5) that equation (1.65) expresses the operation of continuous convolution. In other words, the input–output relation of a linear, time-invariant system is a convolution.

Real systems are seldom truly linear and time-invariant. No system can produce unlimited output if we continuously increase the stimulus. We get saturation effects. Also, previous stimuli do change the system's response to subsequent input. For example, if the system has already reached saturation, then any later input would produce no additional response at all. Thus, linear, time-invariant systems are no more than idealised, though very useful, models of the real world.

1.6.2 LINEAR OPERATORS

Equations (1.61) and (1.62) are not the only possible types of linear operators. Differentiation, for example, is a linear operation, since the derivative of the sum of two functions is the sum of their derivatives. Because of this, mathematical discussions of linear operators always try to be abstract and general, in order to cover all possible linear operators. While we would not go quite so far, it would be useful to remember that almost any result we derive for one type of linear operator has an analog for the other types. Both the discrete and the continuous form defined by equations (1.61) and (1.62) will be referred to in the following. For convenience we will define A as the linear operator and the vector or function it operates on, the **operand**, simply as x. Thus, either equation (1.61) or (1.62) may be written as $y = \mathbf{A}x$.

The right-hand sides of equations (1.59) and (1.60) are called **inner products**.

They produce scalars from two functions or vectors. The inner product of two vectors x and y is

$$(x,y) = \sum_i x_i^* y_i \qquad (1.66)$$

and that between two functions is $\int x(t)^* y(t) dt$ where * indicates the complex conjugate. When x and y are real the complex conjugation has no effect. We also see that equations (1.61) and (1.62) form collections of inner products. It is also clear that, expanding x(t) as an orthogonal series, say

$$x(t) = \sum_i y_i(t) X_i$$

requires evaluation of the inner product of x(t) with each orthonormal function $y_i(t)$

$$X_i = \int y_i(t)^*(t) dt$$

Let us also look at the inner product of a vector or function with itself

$$(x,x) = \sum_i |x_i|^2 \qquad \text{or} \qquad \int |x(t)|^2 dt$$

This is always real and non-negative. The square root of (x,x) is called the **norm** of x, denoted as $\|x\|$. Clearly, $\|x\|$ is zero only when x_i or x(t) is zero for every i or t. Consequently, the norm indicates how significant a quantity a vector or a function is.

Now let us consider applying two linear operators in succession. Suppose

$$y(t) = \int A(t,t')x(t')dt'$$

and

$$z(t) = \int B(t,t')y(t')dt' = \int\int B(t,t')A(t',t'')dt''dt'$$

The overall effect is that of a new operator $C(t,t'')$

$$C(t,t'') = \int B(t,t')A(t',t'')dt' \qquad (1.67)$$

which produces $z(t)$ when operated on $x(t'')$. B and A each contain a collection of functions used to take inner products with other functions. In equation (1.67) however, we are producing a new collection of functions in C by taking inner product of each function in B with each function in A. Similarly, in the discrete case, we could produce

$$Z_k = \sum_i C_{ik}x_i$$

with

$$C_{ik} = \sum_h B_{hk}A_{hi}$$

A particular operator, called the **identity operator** and denoted generally as **I**, leaves its operand unchanged. For equation (1.62), the identity operator is $\delta(t-t')$, since integrating this with $x(t')$ just picks out its value at t, giving $y(t) = x(t)$. For equation (1.61) the identity operator is δ_{ik}. More concisely, we write $y = \mathbf{I}x = x$. The table of numbers formed by δ_{ik} is called the **unit matrix**. It has 1's on the diagonal, as by definition $\delta_{ii} = 1$ for any i; all the elements off the diagonal are 0, $\delta_{ik} = 0$ for $i \neq k$. Obviously, insertion of an identity operator anywhere in an equation does not change anything. Thus $\mathbf{IA}x = \mathbf{A}x$ and $\mathbf{AI}x = \mathbf{A}x$. As a result, we can write $\mathbf{IA} = \mathbf{A}$ and $\mathbf{AI} = \mathbf{A}$.

For certain operators **A** we can find **B** such that applying them in succession leaves the operand unchanged. In other words

$$\int B(t,t')A(t't'')dt' = \delta(t-t''), \text{ or } \mathbf{BA} = \mathbf{I} \qquad (1.68)$$

B is called the **inverse** of **A**. Note that not every operator has an inverse. Those that do are called **non-singular** operators; others are called singular operators. We could say that non-singular operators preserve all the information contained in an operand, so that the operand can be recovered. Singular operators destroy part of the information, making recovery impossible. We write the inverse of **A** as $\mathbf{A}^{-1}$. Note that in addition to $\mathbf{A}^{-1}$. **A** $= \mathbf{I}$ we also have $\mathbf{AA}^{-1} = \mathbf{I}$. This is easily shown by applying A to both sides of the first identity, giving $\mathbf{AA}^{-1}\mathbf{A} = \mathbf{A}$, which shows that $\mathbf{AA}^{-1}$ is an operator that leaves A unchanged. In general, however, operator equations cannot be changed in order of precedence, i.e., **AB** is not necessarily equal to **BA**.

An operator is **symmetric** if A(s,t) = A(t,s), or $A_{ik} = A_{ki}$. A is **Hermitian** if A(s,t) = A(t,s)* or $A_{ik} = A_{ki}*$. If **A** is real, then being Hermitian is the same as being symmetric.

Linear operators have wider occurrences than just as models for physical systems. For example, a Fourier transformation is a linear operation. It converts a time function x(t) into a vector X_i. Its inverse is just the Fourier series. The operator is non-singular as it has an inverse.

1.6.3   EIGENVECTORS AND EIGENVALUES

Given a linear operator **A**, we can always find a set of vectors of functions x_i such that

$$Ax_i = \lambda_i x_i \text{ or } \int A(t,s)x_i(s)ds = \lambda_i x_i(t) \tag{1.69}$$

In other words, **A** operating on x_i reproduces x_i times some constant, rather than a different vector. Of course, when **A** is the identity operator this holds for any x. When A is something else, it is nevertheless possible to find a series of x values satisfying equation (1.69). We call x_i the i'th **eigenvector** (or sometimes eigenfunction) and **A**, and λ_i is the i'th **eigenvalue**.

An important property of Hermitian (which includes real and symmetric) operators is that eigenvectors corresponding to different eigenvalues are orthogonal. We have

$$\int x_k(t)^*[\lambda_i x_i(t)]dt = \int x_k(t)^*A(t,s)x_i(s)dsdt$$

$$= [\int x_k(t)A(t,s)^*x_i(s)^*dsdt]^*, \tag{1.70}$$

where we have moved the conjugation sign outside. (Conjugating anything twice leaves it unchanged.) **A** is Hermitian, so that A(t,s)*=A(s,t). We now integrate over t first. Since x_k is an eigenfunction, its integral with A(s,t) produces $\lambda_k x_k(s)$. Thus the above becomes

$$\int x_k(t)^*[\lambda_i x_i(t)]dt = [\lambda_k x_k(s)x_i(s)^*ds]^* = \lambda_k\int x_k(s)^*x_i(s)ds$$

Or $\tag{1.71}$

$$(\lambda_i - \lambda_k)\int x_k(t)^*x_i(t)dt = 0$$

Since λ_i and λ_k are different, we have that $\int x_k(t)^*x_i(t)dt = 0$. It can be shown that the eigenvalues of a Hermitian operator are all real. If $x_i(t)$ is an eigenfunction of **A**, then so is $\propto x_i(t)$ for any constant $\propto$. By choosing $\propto$ appropriately, we can produce a normalised eigenfunction, i.e., $\int |x_i(t)|^2 dt = 1$ giving an orthonormal set. From this it would be possible to approximate other functions as linear combinations of the eigenfunctions. This provides an orthogonal series, analogous to the Fourier series. The set may not be complete, however, and there may exist functions which cannot be expressed exactly as such orthogonal series.

Given a set of orthogonal functions, we can in theory form all their possible linear combinations. These together form what we call a **linear space**. To conclude this discussion of linear operators we will show the relationship between a linear operator and a matrix. In the linear space each function y(t) is a linear combination of the basic set x: viz. $y(t) = \Sigma Y_i x_i(t)$. Thus, each function in the space corresponds to a **vector Y**. Taking the inner product of two functions, for example, gives

$$\int z(t)^* y(t) dt = \int \sum_i Y_i x_i(t) \sum_k Z_k^* x_k(t) dt$$

$$= \sum_{i,k} Y_i Z_k^* \delta_{ik} = \sum_i Z_i^* Y_i, \qquad (1.72)$$

which is the inner product of these two **vectors** corresponding to y and z. Let us also consider a linear operator **B** which converts function y(t) into another function, say z(t). Both can be expanded in terms of the basic functions x. The eigenfunctions of **B** are of course different from the x's, which are eigenfunctions of **A**. Let us say that **B** operating on $x_i(t)$ converts it into $b_i(t)$, which can again be expressed as a linear combination of x's

$$b_i(t) = \sum_k b_{ik} x_k(t)$$

Now we have

$$z = By = \sum_i Y_i B x_i = \sum_{i,k} Y_i b_{ik} x_k \qquad (1.73)$$

Since $z = Z_k x_k$ we have

$$Z_k = \sum_i b_{ik} Y_i \qquad (1.74)$$

This is the multiplication of a **matrix** into the vector representing y to produce a vector representing z. Thus, the matrix formed by the value b_{ik} represents the operator **B**, even though it was originally defined in a non-matrix way.

This example demonstrates the power of linear mathematics: every linear operator has a corresponding matrix which behaves like the operator in all the essential mathematical relations; two linear operators corresponding to the same matrix would behave in the same way, even though they have been defined in very different forms.

References

1. SCHUSTER, A. On the investigation of hidden periodicities. *Terr. Mag.*, **3**, 13, 1898.
2. SCHUSTER, A. The periodogram of magnetic declination. *Trans. Camb. Phil. Soc.*, **18**, 107, 1900.
3. WHITTAKER, E. T. and ROBINSON, G. *The Calculus of Observations*. Blackie & Sons, London, 1924.
4. SLUTSKY, E. The summation of random causes as the source of cyclic processes. *Econometrica, **5**, 105, 1937.*
5. WIENER, N. *The Extrapolation, Interpolation and Smoothing of Stationary Time Series*. MIT Press, Cambridge and New York, 1949.
6. TUKEY, J. W. The sampling theory of power-spectrum estimates. Symposium in application of autocorrelation analysis to physical problems. *U.S. Office Naval Research*, NAVEXOS-P-735, 1950.
7. BARTLETT, M. S. Periodogram analysis and continuous spectra. *Biometrika*, **37**, 1–16, 1950.
8. GODFREY, M. D. An exploratory study of the biospectrum of economic time series. *Appl. Stat.* C, **14**, 1, 48–69, 1965.
9. COOLEY, J. W., and TUKEY, J. W. An algorithm for the machine calculation of Fourier series. *Math. Comp.*, **19**, 297, 1965.
10. STOCKHAM, T. G. High-speed convolution and correlation. *AFIPS Proc. Spring Joint Comp. Conf.*, **28**, 229–33, 1966.
11. HELMS, H. Non-recursive digital filter design method for achieving specifications of frequency response. *IEEE Trans. Audio and Electroacoust.*, **AU-16**, 1968.
12. NOLL, A. M. The cepstrum and some close relatives. *Signal Processing* (eds. Griffiths, Stocklin and Schooneveld), 11–22, Academic Press, 1973.
13. BURGH, J. P. *Maximum Entropy Analysis*. Ph.D. Thesis, Stanford University, Stanford, California, U.S.A. 1973.
14. PISARENKO, V. F, The retrieval of harmonics from a covariance function. *Geophysics. J. R. Astron Soc.*, 347–66, 1973.
15. PRIESTLEY, M. B. Evolutionary spectra and non-stationary processes. *J. Roy. Stat. Soc.*, B, **27**, 204–37, 1965.
16. BENDAT, J. S., and PIERSOL, A. G. *Random Data: Analysis and Measurement Procedures*. John Wiley, New York, 1971.
17. FRIEDLAND, B. Sampled data control systems containing periodically varying members. *Dept. Elec. Eng. Tech. Rep. Columbia University*, N.Y., T 39/B, Nov. 1959.
18. ROBERTS, J. B., and GASTER, M. Rapid estimation of spectra from irregularly sampled records. *IEE Colloquium on Random Signal Analysis*, IEE London, April 1977.
19. SHANNON, C. E. A mathematical theory of communication. *Bell System Tech. J*, **27**, 623–56, 1948.
20. WIDROW, B. Statistical analysis of amplitude quantised sampled data systems. *IEEE Trans. (Appl. and Ind.)*, **52**, 555, Jan. 1961.

ADDITIONAL REFERENCES

BLACKMAN, R. B., and TUKEY, J. W. *The Measurement of Power Spectra*. Dover, New York, 1959.

JENKINS, G. M., and WATTS, D. G. *Spectral Analysis*. Holden-Day Inc., San Francisco, 1969.

LYNN, P. A. *An Introduction to the Analysis and Processing of Signals*. John Wiley, New York, 1973.

GRIFFITHS, J. W. (Ed.). *Signal Processing*. Academic Press, 1973.

RABINER, L. R., and RADER, C. M. *Digital Signal Processing*. New York, IEEE Press, 1972.

RABINER, L. R., and GOLD, B. *Theory and Application of Digital Signal Processing*. Prentice-Hall, Englewood Cliffs, N.J., 1975.

OPPENHEIM, A. V., and SCHAFER, R. W. *Digital Signal Processing*. Prentice-Hall, Englewood Cliffs, 1975.

SCHWARTS, M., and SHAW, I. *Signal Processing: Discrete Spectral Analysis, Detection and Estimation*. McGraw-Hill, New York, 1975.

Chapter 2

Random Processes

2.1 Introduction

In this chapter we introduce the reader to some of the statistical concepts one requires to gain a proper understanding of signal analysis. Earlier, we met the term 'random signals'. Now one might ask: why would 'random' signals be useful? How can 'random' signals carry information? To answer these questions, we must first explain that, by 'randomness' we mean 'unpredictability'. The important point is that to be able to carry information a signal must appear to be unpredictable to the recipient: if he could predict what a signal is going to be before receiving it, then there is no point in sending the signal. Similarly, if the recipient could predict a future part of the signal from its past values, then that part would simply be redundant. The same applies to the measurement of any physical quantity: to tell us something 'new', each new measured value must be non-predictable from whatever past knowledge we might possess, or whatever past measurements we might have made.

In short, signals must be structured in such a way that past values do not completely determine future values. This does not mean that they are *un*structured, or have arbitrary structure. If nothing else, the structure of any signal is inevitably restricted by the system that produces it or transmits it. Thus, even though a recipient cannot predict future signal values, he may nevertheless know something about their general properties, e.g., they are within the range $[-a, a]$, they do not fluctuate faster than 20,000 Hz, etc.

There is another source of unpredictability: the signals we measure or receive are often contaminated by **noise**, which is the general name for unpredictable disturbances interfering with the signals. Whereas the unpredictability of signals is an essential part of their information carrying capacity, the unpredictability of noise is detrimental to our job because it obscures the signal being transmitted. However, though signal and noise are both unpredictable, they may be different in their general properties, such that we can, to some extent, separate them and recover most, if not all, of the information in the signal.

We are also faced with a difficulty concerning the duration of the signal presented for analysis. Consider first a single record taken from the system. This could be analysed to produce various statistical properties such as mean value, probability distribution, power spectral density, etc. The accuracies of these estimations will depend on the fidelity of the acquisition methods, the length of the record and the method of analysis. Even if we succeed in carrying out the acquisition and analysis processes to a high degree of accuracy we are still left with the uncertainty arising from the limited length of the record since the statistical properties of the process only apply to this particular period in time. From these results we may wish to determine the accuracy of this description and also to extrapolate the measurement into a future time for production of the expected behaviour of the system.

Where the signal is periodic in form then a limited duration of record is adequate to provide a highly descriptive accuracy of estimation. Random signals on the other hand, demand unlimited length of record for exact estimation of their properties and only a limited accuracy is possible in a practical case. It is for this reason that multiple sample records, taken at different periods, are often acquired so that joint analysis of these can enable a higher statistical accuracy to be obtained.

As things turned out, it is in statistics that we find the mathematical tools for overcoming these problems. Statistical techniques allow us to say: 'I do not know what will occur, but I know that what occurs must be one of these events, and each event has so much chance of occurring. On the basis of this knowledge, I make the following decision...' or 'These values vary randomly, but this combination of the values has a more stable behaviour, and I can be sure that it would be pretty close to such and such.'

With the above preamble, we are now ready to begin the study of random variables and random processes. For obvious reasons, the amount of space we can devote to the topic is rather limited. The reader should consult introductory texts on statistics and probability (e.g., [1] [2]) for detailed discussion, particularly for omitted proofs of results obtained in this chapter.

A random variable is something whose value cannot be predicted, except to the extent that it may take on one out of a set of values, each with some **probability**. By probability we mean a number that is associated with a future event and whose size indicates how likely it is that the event will actually occur. To take a most familiar example, coin flipping, we have that the probability of turning a head and that of getting a tail are both 1/2, meaning that heads and tails are equally likely. Let us say, assign value 1 to heads and 0 to tails, then the number we generate by coin flipping is a random variable: we cannot predict its size beforehand, except to the extent that it can take on either 0 or 1, each with probability 1/2.

Now suppose we flip our coin once every second and write down a sequence of 1's and 0's according to the results of the toss, we produce a function of time whose value varies randomly. Functions like that are called **random processes**. As referred to earlier an essential feature of random processes is that future values cannot be expressed as mathematical functions of past values. Thus, a random process can never be a periodic function of time, as such a function has future values which are just repetitions of the past; nor indeed any simple mathematical function altogether.

With real life quantities things are of course not so clear cut. Is, for example, the trajectory of a rocket a well defined mathematical function? Or is it a random process? We can obviously compute an approximate trajectory from the initial velocity of the rocket, its angle of departure, wind velocity, etc. To obtain a more accurate trajectory we would have to include information concerning unpredictable disturbances such as air turbulence, uneven firing, etc., which means that the speed, position and orientation of the rocket must be treated as random variables. Taking these factors into account we could then predict the final landing position of the rocket within a given margin.

While the past of a random process does not determine its value at time t, it may tell us something about what kind of values may turn up and with what chances. Take, for example, the random draw from a bag containing M black balls and N white balls. If, say, we have already taken m black balls and n white balls out of the bag in the previous m+n draws, these do not decide what the colour of the next drawn ball would be. However, the past information does tell us that there are now $M-m$ black balls and $N-n$ white balls remaining in the bag, so that the chance of drawing white next time is $(N-n)/(N+M-m-n)$. Thus, the past of a random process does determine the **probability distribution** of future values, though not the values themselves.

This brings us to the concepts of *a priori* and *a posteriori* probabilities. Let us look again at the example of ball drawings. Suppose we started with a bag containing two white balls and two black balls, and have already taken one ball out but did not look at its colour. What is the chance of getting a white ball next time? Since we have not seen the drawn ball, there is chance 1/2 that there are now two white balls and one black ball left in the bag, and also chance 1/2 that there are two black balls and one white ball. The former would mean that there is chance 2/3 of drawing white next time; in the latter case the chance is 1/3. The chance for drawing black is exactly the reverse. Obviously, since we do not know which case is the correct one, we must say that we are as likely to draw white as black. On the other hand, if we already know that the previous draw was black, then the chance for drawing white next time must be 2/3.

The above example shows that the probability of an event varies with our knowledge about other events. The probability for an event when we have no knowledge about some other event is called the *a priori* probability; that when we do have this knowledge is called the *a posteriori* probability. In the above example, the *a priori* probability for drawing white is 1/2, and the *a posteriori* probability knowing that the previous ball was black is 2/3; for drawing black, the chances are 1/2 and 1/3 respectively. It should be pointed out, however, that the fact that probability of an event varies with our knowledge does not imply that our knowing or not knowing can somehow influence the event. Quite obviously, our looking or not looking at the previous ball cannot influence the next draw. What it does affect is our **ability to make predictions about the next draw**. In other words, probability is just a way of summarising our knowledge about an event.

In this sense we can apply probability to past events as well as future events. Suppose we have just tossed a coin but have not looked at the result. There is nothing **random** about the coin now: it is lying there; its state, whether head or tail, is now fixed. Yet, until we actually look at it we would not know whether to expect a head or a tail: we are equally likely to see either, and the probability for each is still 1/2. In a similar sense, we can assign probability to the colour of the previous ball drawn from the bag using our knowledge of the next ball. Let us say that we have drawn two balls from a bag originally containing two black and two white balls. We do not know the colour of the first ball, but the second ball is black. Now there are six possible cases. Denoting the two white balls as w1, w2 and the two black balls as b1, b2, the following six possible draw combinations would have a black ball on the second draw.

w1,b1; w2,b2; w1,b2; w2,b2; b1,b2; b2,b1

We see that, in four cases the first ball was white; in two the first ball was black. We do not know which had actually occurred. All we know is that one of them did because we know that the second ball was black. So there is chance 4/6 = 2/3 that we had drawn white previously, chance 2/6 = 1/3 that we had drawn black. These are the *a posteriori* probability distributions for the colour of the first ball knowing the colour of the second ball. The *a priori* probability for either colour is of course 1/2. We have thus seen that, in talking about *a priori* and *a posteriori* probabilities, the timing of the events is not material. What is important is our knowledge about them.

Let $P(A)$ and $P(B)$ be the *a priori* probabilities of events A and B, and $P(B/A)$ the *a posteriori* probability for B knowing that A has actually occurred. It may be that A cannot affect B, or that it can have an effect but we do not know enough about their relation, then knowing whether A has occurred does not help us in making predictions about B. In this case there is no difference between the *a priori* and *a posteriori* probabilities, and we have $P(B/A) = P(B)$. Such two events are said to be **independent**. For example, in tossing a coin a number of times we do not expect the results of previous tosses to affect future tosses. Similarly, if after we draw a ball from the bag, we record its colour but then put it back, then the results of different draws would also be independent. But if the balls drawn out stay out then the draws are no longer independent, and certainly the *a priori* and *a posteriori* probabilities are unequal. In the example, $P(B) = 1/2$ while $P(B/A) = 2/3$.

Now let us introduce the concept of **joint probability**, $P(A,B)$, the chance of both A and B occurring. This is an *a priori* probability because it does not depend on our knowledge of other events. We shall state without proof that

$$P(A,B) = P(B/A)P(A) = P(A/B)P(B) \qquad (2.1)$$

To illustrate, consider again the ball draw problem. Let event A be our getting a white ball on the first draw, and B be our getting black on the second draw. There are four ways in which this can happen

$$w1,b1; \quad w2,b1; \quad w1,b2; \quad w2,b2$$

but there are eight ways in which A and B do *not* both occur

$$w1,w2; \quad w2,w1; \quad b1,b2; \quad b2,b1; \quad b1,w1; \quad b2,w1; \quad b1,w2; \quad b2,w2$$

Consequently, the probability for seeing both events is $4/12 = 1/3$, and the probability for not seeing both events is $8/12 = 2/3$. We see that this agrees with what we get by the use of equation (2.1)

$$P(A,B) = (2/3)(1/2) = 1/3$$

We could look at equation (2.1) this way: in the absence of any other knowledge, A has chance $P(A)$ of occurring. Now knowing that A has actually occurred, the chance of B occurring is $P(B/A)$. The chance of both occurring must increase with each of these probabilities, and obviously equation (2.1) satisfies this relation. We might also note that, knowing $P(B/A)$, $P(A)$ and $P(B)$, we can derive $P(A/B)$ from equation (2.1). The significance of this is as follows: $P(A)$, $P(B)$ and $P(B/A)$ can usually be derived by analysing a situation carefully, with $P(B/A)$ derivable from the effect of past on future. We can then obtain $P(A/B)$ and use it to make guesses about what happened in the past using knowledge about its effects. For example, in communication problems the signal received, B, may be different from the signal sent, A, because of noise. By analysing the general behaviour of the noise we determine $P(B/A)$, the chance of getting distorted message B when A was sent. We then derive $P(A/B)$, and use it to determine how likely it is that A had been sent even though B was actually received.

It is clear from equation (2.1) that, if events A and B are independent then

$$P(A,B) = P(A)P(B) \qquad (2.2)$$

This does not hold for the ball drawing case as the events are not independent. Thus, $P(A,B) = 1/6$ whereas $P(A)P(B) = 1/4$. On the other hand, successive tosses of a coin are independent, so that the chance of getting N heads in a row is just $(1/2)^N$, as is well known.

By the very way it is defined, probability is always positive or zero. Negative probability has no sensible meaning. Also, the sum of the probabilities of all the possible events is 1, i.e. if there are M possible events E_i, $i = 1, 2, \ldots, M$, then

$$\sum_{i=1}^{M} P(E_i) = 1 \qquad (2.3)$$

If the situation is such that there is only one possible event, then this event must have probability 1. Events known to be impossible to occur have probability 0.

2.2 Some Familiar Statistical Properties

The reader is probably already familiar with the following collection of statistical quantities, being averages related to a time function x(t), defined as they usually appear in practical applications. In the next section we shall define the same quantities in terms of probability, and then explain in Section 2.6 the connection between the two sets of definitions. The present section is provided as a convenient reference point.

Mean value. The average or mean value of continuous data is

$$\bar{x} = \frac{1}{T} \int_{0}^{T} x(t)dt \qquad (2.4)$$

The corresponding value for a discrete set of N measurements, x_i, is

$$\bar{x} = \frac{1}{N} \sum_{i=1}^{N} x_i = \sum_{i=1}^{N} \frac{x_i}{N} \qquad (2.5)$$

Mean-square value. A single figure describing the intensity of random data is the mean-square value. This is the average of the squared values of the time-history record, which for continuous data is

$$\overline{x^2} = \frac{1}{T} \int_{0}^{T} x^2(t)dt \qquad (2.6)$$

and for discrete data

$$\overline{x^2} = \frac{1}{N} \sum_{i=1}^{N} (x_i)^2 \qquad (2.7)$$

Root mean-square (rms) value. The positive root of the mean-square value is called the rms value of the signal, i.e. $(\overline{x^2})^{1/2}$. Note that generally this does not equal $\overline{x}$.

Variance. A signal can consist of a slowly variable trend or low-frequency component upon which are superimposed rapidly changing higher-frequency components. In such a case the mean value will represent the average base-level of the signal about which it is fluctuating at the higher rate. A measure of the scatter of a signal about its average mean value is given by the variance of the signal. It is described for continuous signals as the mean-square value about the mean, viz.

$$V(x) \;=\; \frac{1}{T} \int_0^T [x(t) - \overline{x}]^2 . dt \qquad (2.8)$$

and for discrete data

$$V(x) \;=\; \frac{1}{N} \sum_{i=1}^{N} (x_i - \overline{x})^2 \qquad (2.9)$$

Standard deviation. In order to specify the instantaneous extent of the deviation of the signal from a mean value we define the deviation as

$$\delta x = x_i - \overline{x} \qquad (2.10)$$

The sum of the deviations for all the data is zero as x is equally likely to be above or below $\overline{x}$, so that it will be necessary to use another method if a single figure is required to identify the deviation value of the entire signal. This is obtained by taking the root of the average of the squares of the instantaneous deviations, since this will always be a positive real number.

$$[\frac{1}{N} \sum_{i=1}^{N} (x_i - \overline{x})^2]^{1/2} \qquad (2.11)$$

This value is called the **standard deviation** of x, and is usually denoted as σx.

Clearly,
$$\sigma x = [V(x)]^{1/2} \qquad (2.12)$$

When x is a zero mean function, i.e., $\overline{x} = 0$, then

$$V(x) = \overline{x^2} \qquad (2.13)$$

and σx equals the rms value of x.

2.3 Probability Distributions and Averages

We are now ready to go into more quantitative discussions of probability.

Suppose x can take any one of a set of values x_i, $i = 1, 2, \ldots M$. The set is called the **ensemble** of x. From this we can define a set of M events, each being the event of x actually taking value x_i. The probability for each event is written as

$$P_x(x_i)$$

Given two random variables x and y, with ensemble sizes of M and N respectively, there would be MN possible events, with x and y each taking one of their permissible values. Each of these events can be assigned a joint probability $P_{xy}(x_i, y_j)$. If we have m random variables, each able to take on one out of M permissible values, then we can define various joint probabilities, the most complicated one being that involving all m variables. There would be M^m possible events, in which each variable takes on one particular value out of the permissible M.

Now let us suppose that x is a function of time, $x(t)$. As t changes, x takes one of the M values but with changing probability for each value, depending on what the past values have been. Thus, $P_x(x_i)$ is a function of time. We write as $p_x^t(x_i)$ the probability that x takes the value x_i at time t, and $p_x^{t,s}$ (x_i, x_k) denotes the joint probability that x takes the value x_i at time t and value x_k at time s. Both are *a priori* probabilities because we have not assumed any knowledge about the values of x at other times. The *a posteriori* probability that x takes value x_i at t knowing that it has taken value x_k at s, which is denoted as $p^{t,s}(x_i/x_k)$, can be found by making use of equation (2.1)

$$p_x^{t,s}(x_i/x_k) = p_x^{t,s}(x_i, x_k) / p_x^s(x_k) \tag{2.14}$$

Other, more complicated types of probabilities can be defined, but are not required for our purpose. We shall show examples of these in Section 2.5. A random process is said to be **time-invariant** or **stationary** if $p_x^t(x_i)$ does not vary with t, and $p_x^{t,s}(x_i, x_k)$ is a function of $(t - s)$ rather than t and s separately.

The second property implies that, no matter what t is, knowing $x(t - \tau)$, its value τ moments earlier, has the same effect on our ability to predict $x(t)$, since $p_x^{t, t-\tau}$ varies only with τ, not with t itself. This concept is very useful in the mathematical analysis of signals. However, there do exist signals that do not possess this stationary property as described in the previous chapter.

Suppose now we define another quantity $z(x)$ in terms of x, then z is also a random quantity as it can take one of the values $z(x_i) = z_i$, each with some probability. If the correspondence between x and z is unique, then the z_i's are different, and z has ensemble size M. If, however, more than one value of x produces the same value of z, then z has an ensemble size of less than M. The probability that z takes any particular value in its ensemble is just the sum of the probabilities of the corresponding values of x. If x is time-invariant, then $p_z^{t,s}$ and p_z^t would satisfy the criteria for time-invariance, and $z(t)$ would thus be time-invariant like $x(t)$.

2.3.1 AVERAGES

Earlier we stated a number of simple statistical properties and their definitions. They were **time averages** because their definition requires integration or summation over time. Some of these properties will now be considered in terms of probability, as **ensemble averages**, because now summation is over the ensemble of x. In short, time averaging is performed over the values of x that **actually** occur, while ensemble averaging takes place over values that are **theoretically** possible. We shall explain their relation in Section 2.6.

We define the **ensemble average** of x as

$$<x> = \sum_{i=1}^{M} P_x(x_i)x_i \qquad (2.15)$$

Applying this to z(x) we have

$$<z> = \sum_{i=1}^{M} P_z(z_i)z_i = \sum_{i=1}^{M} P_x(x_i)z(x_i) \qquad (2.16)$$

$<x>$ and $<z>$ may vary with time, except that if x is time-invariant then p_x^i does not change with t, so that $<x>$ and $<z>$ would also be constant. $<x>$ is also called the *mean* of x.

Let us take $z = x - A$, A being some constant. We have

$$<z> = \sum_{i=1}^{M} P_x(x_i)(x_i - A) = \sum_{i=1}^{M} P_x(x_i)x_i - A\sum_{i=1}^{M} P_x(x_i) \qquad (2.17)$$

$$= <x> - A$$

If we choose $A = <x>$, then $<z> = 0$. In other words, we can produce a random variable with zero mean from one with non-zero mean by a simple subtraction. Note, however, that this cannot be done if $<x>$ varies with time.

Now let us look at $z(x) = x^2$. The average of this variable is called the **mean square** of x

$$<x^2> = \sum_{i=1}^{M} P_x(x_i)(x_i)^2 \qquad (2.18)$$

The mean square of a random variable indicates how significant it is on average. Thus, $<x^2> = 0$ implies that x is identically zero for all purposes: every term in the sum in equation (2.18) is non-negative, so that the sum can be zero only when every term is zero, which implies that for every i either $x_i = 0$ or $P_x(x_i) = 0$, so that x has zero chance of being non-zero. In comparison, $<x> = 0$ would not necessarily mean that x is identically zero, only that its chance of being positive is as likely as being negative.

As we said earlier, the random variable $x-<x>$ has zero mean. Its mean square is called the **variance**, also **dispersion**, also **mean-square deviation** of x: and has the special symbol $V(x)$

$$V(x) = \sum_{i=1}^{M} P_x(x_i)(x-<x>)^2 = <(x-<x>)^2>$$

This average measures the deviation of x from its mean value. If $V(x) = 0$, then we know that x has zero probability of being different from its mean-value. $[V(x)]^{1/2}$, usually denoted as σ_x, is called the **standard deviation** of x, being in a sense the average amount by which x differs from $<x>$. A useful relation is

$$V(x) = <x^2> - 2<x<x>> + <<x^2>> = <x^2> - <x>^2$$

$$(2.19)$$

2.3.2 AVERAGE OF TWO VARIABLES

Given two random variables x and y, with joint probability distribution $P_{xy}(x_i,y_k)$, we can define some averages more complex than the above. The following is called the **covariance** of x and y

$$C(x,y) = <(x-<x>)(y-<y>)> \qquad (2.20)$$

More explicitly this is

$$C(x,y) = <xy>-<x<y>>-<y<x>>+<x><y> = <xy>-<x><y>$$

$$(2.21)$$

with
$$<xy> = \sum_{i,k=1}^{M} P_{xy}(x_i,y_k)x_iy_k \qquad (2.22)$$

An important relation is that

$$<x^2><y^2> \geq <xy>^2 \qquad (2.23)$$

To show this, let us form new random variables $\xi = x/<x^2>^{1/2}$ and $n = y/<y^2>^{1/2}$

Clearly,
$$<\xi^2> = <n^2> = 1$$

Now

$$0 \leq <(\xi+n)^2> = <\xi^2>+<n^2>+2<\xi n>$$

and
$$0 \leq <(\xi-n)^2> = <\xi^2>+<n^2>-2<\xi n>$$

which gives $\qquad -1 \leqslant <\xi n> \leqslant 1$

or $\qquad <xy>^2/<x^2><y^2> \leqslant 1$

This proves equation (2.23). In a similar manner we can prove

$$<x^2>+<y^2> \geqslant 2<xy> \qquad (2.24)$$

Equations (2.23) and (2.24) apply to any pair of random variables, including variables $x-<x>$ and $y-<y>$, which turn these into

$$V(x)V(y) \geqslant [C(x,y)]^2 \qquad (2.25)$$

and $\qquad V(x)+V(y) \geqslant 2C(x,y) \qquad (2.26)$

Taking square roots in equation (2.25) gives from equation (2.12)

$$\sigma_x\sigma_y \geqslant | C(x,y)| \qquad (2.27)$$

The quantity $\rho(x,y) = C(x,y)/\sigma_x\sigma_y$ is called the **correlation coefficient** of x and y. It follows from equation (2.27) that

$$| \rho(x,y)| \leqslant 1 \qquad (2.28)$$

We can easily show that if x and y are independent, i.e. $P_{xy}(x_i,y_k) = P_x(x_i)P_y(y_k)$, then

$$C(x,y) = <x-<x>><y-<y>> = 0$$

so that $\rho(x,y) = 0$. On the other hand, if y is linearly related to x, $y = Ax+B$ for constants A and B, then

$$C(x,y) = A<(x-<x>)^2> = \pm\sigma_x\sigma_y \quad (+ \text{ if A is positive}, - \text{ if A is negative})$$

so that $\rho(x,y) = \pm 1$. Roughly, $\rho(x,y)$ indicates how 'closely related' x and y are: it is 0 if they are independent, ± 1 if they are linearly related, so that we can judge how close we are to either extreme by examining ρ.

These special averages of two variables will be considered in more detail in Chapter 5.

2.3.3 EXAMPLES OF PROBABILITY DISTRIBUTIONS

Binomial distribution. Let us consider a random variable x generated in the following manner: Take a biased (unfair) coin which turns heads with probability P and tails with probability $1-P$. Flip it N times, and the total number of heads is taken as the value of x. Clearly, x is an integer between 0 and N, but we cannot predict which it will be. If we repeat the operation many times, x will take the permissible values in its ensemble with some probability distribution. It is a simple matter to find its probability distribution. There are $N!/[x!(N-x)!]$ different combinations in which x heads and $N-x$ tails show up in a run of N tosses. Further, the probability for each of these combinations is $P^x(1-P)^{N-x}$: the chance of having x heads is P^x and that of having $(N-x)$ tails is $(1-P)^{N-x}$. Hence, the total probability of all these combinations is just

$$P_x(x) = \frac{N!}{x!(N-x)!} P^x(1-P)^{N-x}, \quad x = 0,1\ldots,N \qquad (2.29)$$

Note that the binomial theorem gives

$$(p+q)^N = \Sigma \frac{N!}{x!(N-x)!} p^x q^{N-x}$$

so that

$$\Sigma P_x(x) = [P+(1-P)]^N = 1$$

What are the averages relating to x? First we have

$$<x> = \sum_{x=0}^{N} \frac{N!}{x!(N-x)!} xP^x(1-P)^{N-x}$$

$$= NP \sum_{x=1}^{N} \frac{(N-1)!}{(x-1)![(N-1)-(x-1)]!} P^{x-1}(1-P)^{(N-1)-(x-1)}$$

$$= NP \sum_{y=0}^{N-1} \frac{(N-1)!}{y!(N-1-y)!} P^y(1-P)^{N-1-y}$$

$$= NP[P+(1-P)]^{N-1} = NP \qquad (2.30)$$

This is of course quite reasonable: The average number of heads is the total number of tosses times the chance for each toss turning a head. But this does not mean that we would get NP heads for every run of N tosses. x fluctuates randomly from run to run, sometimes exceeding NP, sometimes falling below that. As we said earlier, the variance of x measures how close x tends to come to its mean value. We can find the variance by making use of equation (2.19)

$$<x^2> = \sum_{x=0}^{N} \frac{N!}{x!(N-x)!} x^2 P^x(1-P)^{N-x}$$

But x^2 is equal to $x + x(x-1)$, so that

$$<x^2> = <x> + N(N-1)P^2 \sum_{x=2}^{N} \frac{(N-2)!}{(x-2)![(N-2)-(x-2)]!} P^{x-2}(1-P)^{(N-2)-(x-2)}$$

The first term is known: the second becomes $N(N-1)P^2 [P+(1-P)]^{N-2}$

$= N(N-1)P^2$ after we put $y = x - 2$ and sum over y. Thus, using equation (2.19) we get

$$V(x) = NP + N(N-1)P^2 - (NP)^2 = NP(1-P) \quad (2.31)$$

and also $\sigma_x = [NP(1-P)]^{1/2}$

Thus, as N goes up $<x>$ increases with N, but σ_x increases only with $\sqrt{N}$. In short, as N becomes large x tends to fall into a small portion of its permissible values as its fluctuation about its mean increases slowly with N.

*Poisson distribution.** A special case of the binomial distribution is the Poisson distribution. This occurs if N is very large, N→∞, but P→o, in such a way that NP, which is just $<x>$ as we saw, remains finite. Let us write NP as λ. Then $P = \lambda/N$. As N→∞ $N!/(N-x)! \to N^x$, so that

$$P_x(x) \to \frac{N^x}{x!} P^x(1-P)^{N-x} = \frac{\lambda^x}{x!} (1-P)^{\lambda/P-x}$$

Take the limit P→0, so that $(1-P)^{\lambda/P-x} \to [(1-P)^{1/(-P)}]^{-\lambda} \to \exp(-\lambda)$ from the definition of exp. We then have

$$P_x(x) \to \frac{\lambda^x}{x!} \exp(-\lambda) \quad (2.32)$$

Note that, instead of having parameters N and P as in the binomial distribution, the Poisson distribution has only one parameter λ, which is, by definition, NP. Note also that we still have

$$\sum_{x=0}^{N} P_x(x) = 1$$

since now N=∞ and the summation is $\exp(-\lambda) \sum_{x=0}^{\infty} \lambda^x/x!$. We recognise that the formula is just $\exp(-\lambda) \exp(\lambda) = 1$

We already know that for the Poisson distribution

$$<x> = NP = \lambda \quad (2.33)$$

Now $$V(x) = NP(1-P)$$

but as $$P\to0, V(x) = NP = \lambda \quad (2.34)$$

We also have $$<x^2> = <x>^2 + V(x) = \lambda + \lambda^2$$

which could have been derived directly from equation (2.32) if we wanted to do that.

Figure 2.1 shows the binomial distribution for $N = 10$ and $P = 0\cdot2$. Figure 2.2(a) shows the Poisson distribution for $\lambda = 2$. The maximum is found

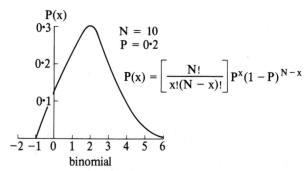

$$P(x) = \left[\frac{N!}{x!(N-x)!}\right]P^x(1-P)^{N-x}$$

Fig. 2.1   The binomial distribution for $N = 10$ and $P = 0\cdot2$

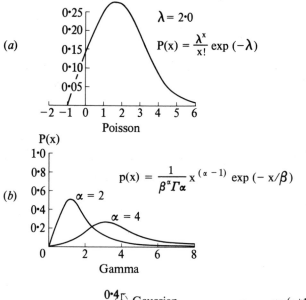

(a)

$$P(x) = \frac{\lambda^x}{x!}\exp(-\lambda)$$

$\lambda = 2\cdot0$

(b)

$$p(x) = \frac{1}{\beta^\alpha \Gamma\alpha}x^{(\alpha-1)}\exp(-x/\beta)$$

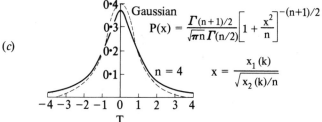

(c)

$$P(x) = \frac{\Gamma(n+1)/2}{\sqrt{\pi n}\,\Gamma(n/2)}\left[1+\frac{x^2}{n}\right]^{-(n+1)/2}$$

$$x = \frac{x_1(k)}{\sqrt{x_2(k)/n}}$$

Fig. 2.2   Some probability distribution functions
(a)   Poisson
(b)   Gamma
(c)   Gaussian and Student

near, but not exactly at, the value of <x>. Both distributions are **unimodal**, i.e., having only one maximum, the position of which is called the **mode** of the distribution.

2.4   Continuous Random Variables

2.4.1   INTRODUCTION

So far we have only considered random variables which are confined to a finite set of values. Now we wish to consider random variables whose values may be anywhere in a continuous domain. We can do this by using the concept of probability density function. This describes the probability that a random variable will take on a value within some defined amplitude range at any instant in time. Thus with reference to Fig. 2.3, the probability that x(t) will take on a value in the range between x_1 and x_2 is given as the ratio of the time that x dwells within this range and the total time of the record, i.e.

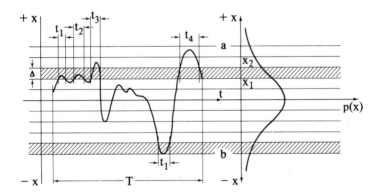

Fig. 2.3   Derivation of probability density function

$$P(x_1 < \xi < x_2) = \lim_{T \to \infty} \frac{t(n)}{T}$$

where

$$t(n) = \Sigma \, t_k$$

n is the level number and k = 1,2,...

Let us now consider this in more rigorous mathematical terms. If we have a random variable confined to the interval [a, b], such that it is equally likely to be any one of the values within the interval then what is its probability distribution? Obviously, $P(x) = 0$ if $x<a$ or $x>b$, but since there is an infinite number of permissible values within the interval, and each value is equally likely, the probability of x attaining a particular value must be infinitesimally small! So we can define a $P(x)$, but it is really not very meaningful.

What we can meaningfully define is a **probability density function**, say $p_x(\xi)$, the size of which indicates how likely x is to be found in the **neighbourhood** of a particular value ξ as illustrated in Fig. 2.3. For the present example, the probability that x is in the sub-interval $[\xi,\xi+\Delta]$ is obviously $\Delta/(b-a)$. Thus, the probability of x being in the neighbourhood of ξ, $[\xi,\xi+d\xi]$, is just $\frac{d\xi}{b-a}$. Thus, we can define function $p_x(\xi)$ such that $p_x(\xi)d\xi$ is the probability of x being in the neighbourhood of ξ. In the present example $p_x(\xi) = 1/(b-a)$.

Now if we want to obtain the probability of x being in the sub-interval $[\xi,\xi+\Delta]$, we can just divide the sub-interval into infinitesimal neighbourhoods and sum all the probabilities

$$P(\xi \leqslant x \leqslant \xi+\Delta) = p_x(\xi)d\xi + p_x(\xi+d\xi)d\xi + \ldots + p_x(\xi+\Delta-d\xi)d\xi$$

This is the definition of an integral

$$P(\xi \leqslant x \leqslant \xi+\Delta) = \int_{\xi}^{\xi+\Delta} p_x(\xi)d\xi \qquad (2.35)$$

Note that for $p_x(\xi) = 1/(b-a)$ we have

$$\int_{\xi}^{\xi+\Delta} p_x(\xi)d\xi = \frac{1}{(b-a)}(\xi+\Delta-\xi) = \frac{\Delta}{b-a}$$

which is of course correct. And

$$\int_{a}^{b} p_x(\xi)d\xi = 1 \qquad (2.36)$$

showing that x must be in the interval [a,b].

Earlier we defined ensemble averages by summation. With continuous domains, averaging is now performed by integration. Thus

$$<x> = \int p_x(\xi)\xi d\xi$$

and
$$<x^2> = \int p_x(\xi)\xi^2 d\xi$$

and so on. With probability density functions of more than one variable, we can define more complex averages, analogous to those shown in Section 2.3.2 in terms of multiple integration.

For theoretical reasons we often encounter the integral of probability density functions

$$P_x(x) = \int_{-\infty}^{x} p_x(\xi)d\xi \qquad (2.37)$$

This is called the **probability distribution function**. It represents the probability that $x(t)$ is inside the interval $-\infty$ to x, i.e.

$$P_x(x) = P(-\infty < x(t) \leq x)$$

For the sake of simplicity we often write $p_x(x)$ as just p(x), and $P_x(x)$ as just P(x), omitting the subsidiary symbols. However, when handling mathematical derivations we require these to avoid confusion.

The reader may well ask this: in practice our measurements have only a finite precision, and the values we get only have a finite number of digits. They can never form a continuous domain. So why do we need the above complications? The reason lies in mathematical convenience, as we illustrate now with an example.

Consider the following problem: Given two random variables x and y, with probability density functions $p_x(\xi)$ and $p_y(\eta)$ respectively, what can we say about the new random variable z = x+y? What, for example, is the probability of z being in the neighbourhood $[\zeta,\zeta+d\zeta]$? We have to sum over all the possible values of x and y such that x+y = ζ, say y ϵ [$\eta,\eta+d\eta$] and at the same time x ϵ [$\zeta-\eta,\zeta-\eta+d\xi$]

$$p_z(\zeta)d\zeta = \Sigma\; p(\zeta-\eta \leq x \leq \zeta-\eta+d\xi, \eta \leq y \leq \eta+d\eta)$$

As x is fixed once we fix y and z, we need only sum (or integrate) over y. Assuming that x and y are statistically independent, so that

$$p(\zeta-\eta \leq x \leq \zeta-\eta+d\xi, \eta \leq y \leq \eta+d\eta) = p(\zeta-\eta \leq x \leq \zeta-\eta+d\xi)p(\eta \leq y \leq \eta+d\eta)$$

$$= p_x(\zeta-\eta)d\xi\; p_y(\eta)d\eta$$

we have $p_z(\zeta)d\zeta = (\int p_x(\zeta-\eta)p_y(\eta)d\eta)d\xi$

As $d\zeta$ cancels $d\xi$ we have

$$p_z(\zeta) = \int p_x(\zeta-\eta)p_y(\eta)d\eta \qquad (2.38)$$

which is a convolution. We see that, once continuous distributions have been defined we can bring to bear familiar mathematical tools on our problem.

To take an example, if x and y are both uniformly distributed in $[-A,A]$ we have $p_x(\zeta-\eta) = p_y(\eta) = 1/2A$. When integrating to find p_z, we have the restriction $-A \leq \zeta-\eta, \eta \leq A$, or

$$\eta \; \epsilon \; [\zeta-A,A] \text{ if } 0 \leq \zeta \leq 2A$$
$$\epsilon \; [-A,\zeta+A] \text{ if } -2A \leq \zeta \leq 0$$

and if $|\zeta| < 2A$ no value of η satisfies the restriction, so that there is nil interval of integration. Thus we have

$$p_z(\zeta) = (2A)^{-2} \int_{\zeta-A}^{A} d\eta = 1/(2A) - \zeta/(2A)^2, \quad 0 \leqslant \zeta \leqslant 2A$$

$$= (2A)^{-2} \int_{-A}^{\zeta+A} d\eta = 1/(2A) + \zeta/(2A)^2, \quad -2A \leqslant \zeta \leqslant 0$$

$$= 0 \text{ if } \zeta > 2A \text{ or } < -2A \tag{2.39}$$

This distribution is shown in Fig. 2.4. We see that z is more likely to be near 0 and less likely to be ±2A.

Suppose we are given x and y having the distribution of Fig. 2.4, we can again find the distribution of z = x + y by a convolution. The result of this is shown in Fig. 2.5. The distribution has the familiar 'bell' shape. Earlier we saw similar bell shapes in the binomial and Poisson distributions. We shall describe next the most important of all the bell shapes: the Gaussian distribution.

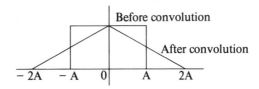

Fig. 2.4   Probability density function (PDF) of the sum of two independent, uniform random variables

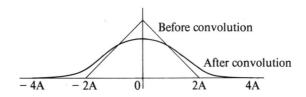

Fig. 2.5   PDF of the sum of four independent, uniform random variables

2.4.2   GAUSSIAN AND RELATED VARIABLES

Normal or Gaussian probability. Gaussian processes are found in many physical situations so that the corresponding probability density function is an important one. This is particularly so since a Gaussian process also results from a linear transformation of a process, which is itself Gaussian. This means that data acquired directly or indirectly from a physical system will still retain its original probability density distribution, if this is Gaussian, despite the distortions obtained in the process of data capture.

The importance of the Gaussian probability density distribution arises as a consequence of the **central limit theorem**, which states that under very general conditions if $x_1, x_x, \ldots, x_n$ are independent random variables whose probability density functions are different (and may be unknown), then the function of the sum of these random variables

$$x = \sum_{k=1}^{n} x_k \qquad (2.40)$$

approaches a Gaussian density distribution as $n \to \infty$. This result applies on the assumption that no particular constituent random variable dominates the behaviour of the summed random variable x. In practice many random processes result from the summation of a large number of independent random inputs, e.g. atmospheric turbulence, pressure from a jet exhaust, vibration of a complex structure and many of the forms of random noise. We will consider first a standardised form of this density function.

Standardising a random variable, x(t), means the process of obtaining a new series of values for z(t) where the mean is zero and the deviation-squared from the mean is unity (i.e. unit variance). We shall see later that this is a common requirement and simplifies subsequent analysis. The requirement is easy to satisfy: if we denote $<x>$ as μ and σ_x as S, then $x(t) - \mu$ must have zero mean. Further, $z(t) = [x(t) - \mu]/S$ must have unit variance because

$$<[z(t)]^2> = \frac{1}{S^2} <[x(t) - \mu]^2> = \frac{1}{S^2}(S^2) \qquad (2.41)$$

Using this normalised standard version of the series the Gaussian, or normal probability density function can be defined as

$$p(z) = \frac{1}{\sqrt{(2\pi)}} \exp(z/2^2) \ (-\infty < z < +\infty) \qquad (2.42)$$

This is a symmetric bell-shaped curve and is shown in Fig. 2.5. It is sufficient to know this function for positive value of z since

$$p(z) = p(-z)$$

The Gaussian probability distribution function is, by definition

$$P(z) = \int_{-\infty}^{z} p(z)dz \qquad (2.43)$$

and defines the probability that $z(t) \leqslant z$.

The width of the Gaussian probability density curve can be expressed in terms of standard deviation S, using equation (2.41). For a zero mean-value of x(t) then the area under a Gaussian probability density curve of width ±2S will be 95 per cent of the total theoretical value thus providing a useful reference point for comparison with other results.

The probability that a random variable z(t) lies between two limits A and B can also be defined as the definite integral of equation (2.42).

$$P(z)_{AB} = \frac{1}{\sqrt{(2\pi)}} \int_B^A \exp(-z/2^2) \, . \, dz \qquad (2.44)$$

Such integrals are difficult to evaluate using exact methods and a difference technique using standard tables is used [3]. These tables generally give area under the standardised normal distribution function (equation 2.43). To use these to obtain $P(z)_{AB}$ we make use of the relationship

$$P(A<z<B) = P(A) - P(B) \qquad (2.45)$$

Taking the difference between the two distribution functions obtained at the limiting values z = A and z = B, a value for the integral is found (Fig. 2.6).

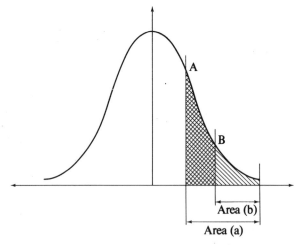

Fig. 2.6   Derivation of $P(z)_{AB}$ = area (a) − area (b)

If the random variable does not have zero mean and unit variance then x = $S_z + \mu$ is the value of the random variable whose probability density function is obtained from equations (2.41) and (2.42) thus

$$p(x) = \frac{1}{S\sqrt{(2\pi)}} \, . \, \exp\left[-\frac{1}{2}\left(\frac{x-\mu}{S}\right)^2\right] \qquad (2.46)$$

which is the usual form of the Gaussian probability density function.

The Gaussian curve shown in Fig. 2.2c is drawn for the normalised standard version where $\mu = 0$ and $S^2 = 1$. Where the mean value and variance depart from these values then we expect some alteration in the position and shape of this curve. For a fixed value of S then a change of mean value, μ, will merely shift the curve laterally without any change of shape (Fig. 2.7). The value of S will, however, considerably affect the shape of the curve. As S becomes larger the curve will tend to flatten and as S becomes smaller the curve will become steeper (Fig. 2.8).

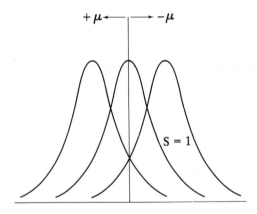

Fig. 2.7   Gaussian curve — shift of mean-value

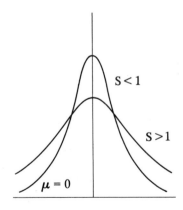

Fig. 2.8   Gaussian curve — change of variance

The fact that x has mean μ, standard deviation S and probability density function p(x) is mathematically expressed as

$$<(x-\mu)^2> = \int_{-\infty}^{\infty} p(x)(x-\mu)^2 dx = S^2 \qquad (2.47)$$

a result which is readily proved. It is also possible to prove the more general result

$$<(x-\mu)^2> = S^n (n-1) (n-3)\ldots 1 \text{ if n is even}$$

$$= 0 \text{ if n is odd} \tag{2.48}$$

For n = 2 this gives result (2.47). For n = 4 we have $<(x-\mu)^4> = 3S^4$, a relation we shall make use of several times later. That we can obtain easily fairly complex averages of Gaussian variables numerically is one of their important properties.

We shall also require the joint density functions of more than one Gaussian variable, say x and y. Suppose they have mean 0 and standard deviation 1, and their correlation coefficient is ρ, then we can immediately write down their joint density function as

$$p(x,y)dxdy = [2\pi(1-\rho^2)]^{-1/2}\exp[(-x/2^2-y/2^2+\rho xy)/(1-\rho^2)]dxdy \tag{2.49}$$

This is in standard form as found in equation (2.42), which converts to (2.46) for non-zero μ and non-zero S. If we require p(x,y) for x and y with other means and standard deviations, we only have to substitute $(x-\mu_x)/S_x$ for x and $(y-\mu_y)/S_y$ for y. Note that if $\rho = 0$, or equivalently, if C(x,y) = 0 then

$$p(x,y) = \frac{1}{2\pi} \exp(-x/2^2)\exp(-y/2^2) = p(x)p(y)$$

In other words, if two Gaussian variables are **uncorrelated** they are also **independent**, which is another special property.

Finally, a useful relation is

$$<xyuv> = <xy><uv> + <xu><yv> + <xv><yu> \tag{2.50}$$

where x, y, u, v are all zero mean Gaussian variables. This relation permits us to evaluate quite complex expressions involving Gaussian variables. Proof of the above results can be found in [4], pp. 47−49, pp. 61−64.

Log-normal. This type of density distribution occurs where the logarithm of the random variable has a Gaussian function. The probability density function of a log-normal distribution is given by

$$p(x) = \frac{1}{xS\sqrt{(2\pi)}} \exp\left[-1/2\left(\frac{\log_e(x)-<\log_e(x)>}{S}\right)^2\right] \tag{2.51}$$

*Gamma density functions.** A number of continuous functions form a particular class of probability density functions known as Gamma functions. A general definition is given by

$$p(x) = \frac{1}{\beta^\alpha \Gamma(\alpha)} . x^{(\alpha - 1)} \exp(-x/\beta) \qquad (2.52)$$

where $\Gamma(\alpha)$ is the value of a gamma function defined by

$$\Gamma(\alpha) = \int_0^\infty x^{(\alpha - 1/2)} \exp(-x) . dx$$

For integer values of α the gamma function can be obtained from

$$\Gamma(\alpha) = (\alpha - 1)!$$

Examples of this function were given in Fig. 2.2(b), from which we see that the distributions are positively skewed with a skew value proportional to $1/\alpha$. This function, and those derived from it, are valuable to describe waiting times for random events, such as the time intervals between successive breakdowns in physical systems, or the time intervals between random 'hits' for a given process. It is closely related to the Poisson distribution, in that they study different aspects of the same process: one describes the number of events per unit time interval, while the other describes the time between successive events. A special gamma density function for $\alpha = 1$ is the exponential function shown in Fig. 2.2 for which

$$p(x) = \frac{1}{\beta} . \exp(-x/\beta)$$

The example shown is for a value of $\beta = 2$.

Rayleigh.* This is another form of skewed gamma function. As with other functions of the gamma class it is extremely non-symmetrical, being zero for negative signal values. The probability density function for a Rayleigh density distribution is given by

$$p(x) = \frac{x}{\gamma^2} . \exp(-1/2(x/\gamma)^2) \qquad (2.53)$$

One use for this distribution is to describe analytically the noise output of an envelope detection system used in radio reception. In such a system the Gaussian noise characteristic of the random noise is modified by the narrow bandwidth so that the probability density of the signal peaks follows a Rayleigh distribution law. In equation (2.53) the factor γ is related to the standard deviation, S of the variable x, by a constant $\gamma = 1 \cdot 526$ S.

2.4.3 SAMPLING DISTRIBUTIONS

The following three probability functions arise initially from statistical inference, where we analyse observed values, also called **sampled values**, of random variables in order to decide whether they are consistent with our hypothesis about their average properties. Depending on the particular type of analysis, one of the following distributions arises:

Chi-squared. It was remarked earlier that when only one random variable is considered then in many physical situations the density function is likely to be Gaussian. Where the number of contributory independent random variables is finite and each random variable is Gaussian, having zero mean and unit variance then we can define a new random variable as

$$\chi_n^2 = z_1^2 + z_2^2 + z_3^2, \ldots, z_n^2 \tag{2.54}$$

here the new random variable χ_n^2 is known as a chi-squared variable having n degrees of freedom.

Degrees of freedom will be defined more specifically later in connection with power spectral density evaluation. For the moment we can consider n to be equal to the finite number of random variables involved in equation (2.54). This form of density distribution is particularly relevant to sampled data where it is permissible to consider each sample as being statistically independent and random. It approaches normal distribution as $n \to \infty$.

Specifically, the probability density function is given by

$$p(x) = \frac{1}{2^{n/2}\Gamma(n/2)} x^{1/2n-1}\exp(-1/2x) \tag{2.55}$$

where $x = \chi_n^2$. Substitution of $\alpha = 1/2n$ and $\beta = 2$ reduces equation (2.55) to the more general case of the gamma density function given by equation (2.52). The set of gamma curves given in Fig. 2.2 are in fact χ^2 functions expressed in terms of a single parameter (degrees of freedom n). This parameter is indicative of the sharpness of the probability selection and is allied to the definition of system performance in the frequency domain. (See Chapter 4.)

The mean and variance of χ_n^2 are easily found, since we know that the z's have zero mean and unit variance, so that

$$<z_i^2> = 1, \quad <z_i^4> = 3$$

according to equation (2.48). Because they are independent we have $C(z_i^2, z_k^2) = 0$ for different i and k. In view of equation (2.23) we know

$$<z_i^2 z_k^2> = <z_i^2><z_k^2> = 1, \quad i \neq k$$

We then find

$$\chi_n^2 = \sum_{i=1}^{n} <z_i^2> = n \qquad (2.56)$$

and

$$(\chi_n^2)^2 = < \sum_{i=1}^{n} z_i^2 \sum_{k=1}^{n} z_k^2 > = \sum_{i=1}^{n} <z_i^4> + \sum_{i \neq k} <z_i^2 z_k^2>$$

$$= 3n + n(n-1) = n^2 + 2n$$

which means

$$V(\chi_n^2) = <(\chi_n^2)^2> - <\chi_n^2>^2 = (n^2+2n) - n^2 = 2n \qquad (2.57)$$

Thus the standard deviation of χ_n^2 is just $\sqrt{(2n)}$.

F-distribution.* A sampling distribution which is used to determine the ratio of the variances of two independent random series is known as the **F-distribution**. If $x_1(k)$ and $x_2(k)$ are statistically independent random variables, each having a χ^2 probability function with n_1 and n_2 degrees of freedom respectively, the new random variable can be defined as

$$x = \frac{x_1(k)n_2}{x_2(k)n_1} \qquad k = 1,2,\ldots$$

The probability density function for x is

$$p(x) = \frac{\Gamma[(n_1+n_2)/2]x^{n/2_1-1}(n_1/n_2)^{n/2_1}}{\Gamma(n/2_1)\Gamma(n/2_2)[1+(n_1/n_2)x]^{(n_1+n_2)/2}} \qquad (2.58)$$

Student-T.* This is also defined for two independent random variables $x_1(k)$ and $x_2(k)$, where the former has a normal distribution and the latter a χ^2 distribution. The new random variable is defined as

$$x = x_1(k)/ \left(\frac{x_2(k)}{n}\right)^{1/2} \qquad (2.59)$$

which is called a student-T variable with n degrees of freedom. The probability distribution is given as

$$p(x) = \frac{\Gamma(n+1)/2}{\Gamma(\frac{1}{2}n)\sqrt{\pi n}} \left(1+ \frac{x^2}{n}\right)^{-1/2(n+1)} \qquad (2.60)$$

This function is plotted in Fig. 2.2(c) with that of the Gaussian function. Both are bell-shaped and symmetrical about the mean. It can be shown that the student-T function approaches the Gaussian function as $n \to \infty$, and differs from it by a negligible amount for n of 50 or more.

2.5   Examples of Random Processes

2.5.1   THE AUTO-CORRELATION FUNCTION

Let us consider two random variables which are values of a random process x at times t and s. We can define various averages from them using their joint probability distribution as shown in Section 2.3.2. The fact that they are values of a random process, however, gives rise to some special properties.

We define the **auto-correlation function**, R(t,s) of the random process as

$$R(t,s) = < x(t)x(s) > = \sum_{i,k} P_x^{t,s}(x_i,x_k)x_ix_k \qquad (2.61)$$

This is an important quantity in that it summarises a great deal of information about the random process, and it arises on various occasions in signal analysis. We shall study this and similar quantities in Chapter 5, though we shall see a few examples in this chapter.

If x is a stationary random process, then $P_x^{t,s}$ is a function of $(t-s)$ only, and we would have

$$R(t,s) = R(t-s) \qquad (2.62)$$

Further, $<x(t)x(s)>$ and $<x(s)x(t)>$ are exactly the same thing, so we have $R(t-s) = R(s-t) = R(-t+s)$. In other words, letting $s-t = \tau$ we can write $R(\tau) = R(-\tau)$, and R is defined as an even function so that we can also write $R(|t-s|)$ if we so wish.

If t = s, R(t−s) = R(0) is just $<x(t)x(t)>$; in other words, R(0) is the same as the **mean-square** of x. Note that this is a constant and does not change with t since x is stationary. Further, in view of equation (2.23) we have

$$[R(0)]^2 = <x(t)^2><x(s)^2> \geq <x(t)x(s)>^2$$

or for any time difference τ

$$R(0) \geq | R(\tau)| \qquad (2.63)$$

When x(t) is an electric signal, its mean-square value gives its average power. We thus have a physical interpretation of R(0). For non-zero differences of time $R(\tau)$ expresses the relation between x at different times, and hence, the time behaviour of x. For example, we may be interested in the covariance between x(t) and x(s), C(x(t),x(s)), in analogy to C(x,y) between two arbitrary random variables x and y. Let us denote this as C(t,s) for convenience. By definition (2.20) this is

$$C(t,s) = <[x(t)-<x(t)>] [x(s)-<x(s)>]> \quad (2.64)$$

$$= <x(t)x(s)> - <x(t)><x(s)>$$

For stationary processes $<x(t)>=<x(s)>=\mu$, and

$$C(t,s) = R(t-s) - \mu^2 \quad (2.65)$$

We can also obtain the correlation coefficient between x(t) and x(s) from R(t−s). By definition (Section 2.3.2)

$$p(x(t),x(s)) = C(x(t),x(s))/\sigma_{x(s)}\sigma_{x(t)} \quad (2.66)$$

For a stationary process $\sigma_{x(s)} = \sigma_{x(t)}$, so that the denominator is just $(\sigma_{x(t)})^2 = V(x(t)) = <[x(t)-\mu]^2> = C(t,t) = R(0)-\mu^2$. Thus, using the notation $\rho(t,s)$ for the correlation coefficient we can write

$$\rho(t,s) = \rho(x(t),x(s)) = [R(t-s)-\mu^2]/[R(0)-\mu^2] \quad (2.67)$$

Further, we saw in equation (2.49) that, once we know the correlation coefficient between two Gaussian variables we can write down their joint probability density function at once. Thus, if we know R(t−s) we also know $p_x^{t,s}$. $\rho(t,s)$ is called the auto-correlation coefficient function, and C(t,s) is the auto-covariance function of x(t).

(We should point out that the above naming system is followed by engineering texts but not by statisticians. The latter call $\rho(t,s)$ the auto-correlation function. They usually do not bother to define R, feeling that ρ and C are the more important quantities.)

In analogy to equation (2.62), we have for stationary processes

$$C(t,s) = C(t-s), \quad \rho(t,s) = \rho(t-s) \quad (2.68)$$

And in analogy to equation (2.63) we have

$$C(0) \geqslant | C(\tau)| , \quad \rho(0) \geqslant | \rho(\tau)| \quad (2.69)$$

Note that $\rho(0) = 1$, as we can see by setting t = s in equation (2.67).

2.5.2 WHITE NOISE

White noise is a special type of Gaussian random process, characterised by the properties $\mu = 0$ and

$$\rho(\tau) = 0 \text{ if } \tau \neq 0 \quad (2.70)$$

Because the process is Gaussian, x(t) and x(s) being uncorrelated implies that they are also independent, or, knowing x at time s does not help us to make predictions about x at other times. Thus, white noise is a random process that exhibits no regular behaviour of any kind, with no discernible relation between its values at different times.

From equation (2.67) we see that equation (2.70) implies

$$R(t-s) = 0 \text{ if } t \neq s$$

Thus for white noise $R(\tau)$ is a function that is zero everywhere except for a spike at the origin. We recognise that this is the Dirac delta function introduced in Chapter 1.

$$R(\tau) = \Lambda\delta(\tau) \qquad (2.71)$$

where Λ is some constant. Later on, we shall see that white noise is physically unrealisable. Rather, it is a useful mathematical model for random interferences which fluctuate rapidly without appreciable regularity. It is usual to denote white noise as $n(t)$.

Now although $n(t)$ shows no regular behaviour, other random processes we define from it may. Take, for example, the weighted integral of $n(t)$

$$x(t) = \int_{-\infty}^{t} w(v)n(v)dv \qquad (2.72)$$

Its auto-correlation function is

$$R(t,s) = \int_{-\infty}^{t} \int_{-\infty}^{s} w(v)w(v')<n(v)n(v')>dvdv'$$

$$= \int_{-\infty}^{t} \int_{-\infty}^{s} w(v)w(v')\delta(v-v')dvdv'$$

Since $\delta(v-v') = 0$ except at $v-v' = 0$, integrating it with any function gives just the value of the integrand at $v = v'$. Thus

$$R(t,s) = \int_{-\infty}^{t} w(v)w(v)dv$$

We see that this is not a function of $t-s$, from which we deduce that $x(t)$ cannot be a stationary random variable.

Now let us consider the new process produced from $n(t)$ by a **moving average**

$$x(t) = \int_{-\infty}^{t} w(t-v)n(v)dv$$

This is a Gaussian process because n is Gaussian and we said earlier 'any linear combination of Gaussian variables is Gaussian'. Its auto-correlation is

$$R(t,s) = \int\limits_{-\infty}^{t} \int\limits_{-\infty}^{s} w(t-v)w(s-v')\delta(v-v')dvdv'$$

Integrating with a δ-function again picks out the value of $w(t-v)w(s-v')$ at $v = v'$. However, if $t>s$ those values of v greater than s would be outside the range of v' and we do not have any v' that can equal such a v.

Similarly, if $t<s$ any v' greater than s does not have a corresponding v.

Thus, only those values of v and v' smaller than **both** t and s will contribute to the integral. Assuming that $t>s$ we must have

$$R(t,s) = \int\limits_{-\infty}^{s} w(t-v)w(s-v)dv = \int\limits_{0}^{\infty} w(v)w(v+t-s)dv$$

where we have made the change in integration variable, replacing v by $s-v$. For $t<s$ we would have

$$R(t,s) = \int\limits_{-\infty}^{t} w(t-v)w(s-v)dv = \int\limits_{0}^{\infty} w(v)w(v+s-t)dv$$

where we have replaced v by $t-v$ as integration variable. We see that in either case $R(t,s)$ is a function of $|t-s|$ only, showing that $x(t)$ is stationary. Generally

$$R(t,s) = \int\limits_{0}^{\infty} w(v)w(v+|t-s|)dv \qquad (2.73)$$

Note that if the moving average is taken over a finite duration only, i.e. if $w(v) = 0$ when v exceeds some finite range Δ, then $R(t,s)$ is zero for any t and s that are separated by more than Δ, since $(v+|t-s|)$ would exceed Δ and $w(v+|t-s|) = 0$ for all v, so that equation (2.73) would give just zero. Thus, usually we have

$$R(t,s) = 0 \text{ if } |t-s| > \Delta$$

for some finite range Δ.

2.5.3 AUTO-REGRESSIVE PROCESSES*

In the last subsection we saw that we can generate stationary random processes from white noise by means of moving averages. In this subsection we discuss another class of stationary processes generated from white noise. These are called **auto-regressive processes** because the value of $x(t)$ is a weighted sum of its previous values plus white noise

$$x(t) = \sum_{i=1}^{m} \alpha_i x(t-i\Delta) + n(t) \qquad (2.74)$$

$$i = 1,2\ldots,m$$

If we take Δ to be infinitesimally small the sum goes into an integral

$$x(t) = \int_{t-m\Delta}^{t} \alpha_i x(t-\tau)d\tau + n(t)$$

This is, however, rather hard to work with in practice as it generates integral equations rather than matrix relations and we usually prefer to work with equation (2.74). Auto-regressive processes are fairly good models of certain real-life systems. For example, the population of a species of animals within a closed region certainly depends on its past values, but it is also affected by random factors like the weather. Such unpredictable disturbances are represented by the $n(t)$ in equation (2.74). In similar ways, we might reasonably think that the present values of stocks determine the trading pattern in the immediate future, and therefore, future prices, with random factors again playing a part.

Let us consider the auto-correlation function of x. First we multiply both sides of equation (2.74) by $x(t)$ and take ensemble average

$$<x(t)^2> = \sum_{i=1}^{m} \alpha_i <x(t)x(t-i\Delta)> + <x(t)n(t)> \qquad (2.75)$$

$n(t)$ is completely independent of the previous values of x as these depend on n at times not equal to t. Thus

$$<x(t)n(t)> = \sum_{i=1}^{m} <\alpha_i x(t-i\Delta)n(t)> + <n(t)^2> = <n(t)^2> = \Lambda$$

so that from the definition for auto-correlation given by equation (2.61) result (2.75) becomes

$$R(0) = \sum_{i=1}^{m} \alpha_i R(i\Delta) + \Lambda \qquad (2.76)$$

We then look at $<x(t)x(t-k\Delta)>$ for $k \geq 1$. We know that $<x(t-k\Delta)z(t)> = 0$, and this gives

$$R(k\Delta) = \sum_{i=1}^{m} \alpha_i <x(t-i\Delta)x(t-k\Delta)> = \sum_{i=1}^{m} \alpha_i R(|k-i|\Delta) \quad (2.77)$$

There is one more relation. Correlating each side of (2.76) with itself and using the fact that n(t) is independent of past values of x we get

$$R(0) = \sum_{i,k=1}^{m} \alpha_i \alpha_k R(|i-k| \Delta) + \Lambda \tag{2.78}$$

Given an auto-regressive model, i.e., given m and α_i for i = 1,2...,m, we can compute R(iΔ) for any i without great difficulty. First we note that if we combine equations (2.78) with (2.77) for k = 1,2,...,m−1, we have m equations in m unknowns, which gives R(kΔ) for k = 0,1,...,m−1. Solving these gives us the first m values of R. We can then use (2.77) to find R(kΔ) for k = m, m+1, etc. However, while numerical values of R are easy to get, it is usually difficult to obtain a general expression.

As example, consider the case of m = 1, or

$$x(t) = x(t-\Delta) + n(t) \tag{2.79}$$

result (2.78) gives

$$R(0) = \alpha^2 R(0) + \Lambda$$

so that

$$R(0) = \Lambda/(1-\alpha^2) \tag{2.80}$$

result (2.77) then gives

$$R(k\Delta) = \alpha R[(k-1)\Delta] \tag{2.81}$$

Applying (2.81) repeatedly we get

$$R(k\Delta) = \alpha^k R(0) = \Lambda \alpha^k/(1-\alpha^2) \tag{2.82}$$

A harder example is m = 2. Equation (2.78) gives

$$R(0) = (\alpha_1^2 + \alpha_2^2)R(0) + 2\alpha_1\alpha_2 R(\Delta) + \Lambda \tag{2.83}$$

while result (2.77) for k = 1 is

$$R(\Delta) = \alpha_1 R(0) + \alpha_2 R(\Delta) \tag{2.84}$$

Solving these we get

$$R(0) = \Lambda(1-\alpha_2)/(1-\alpha_2-\alpha_1^2 + \alpha_2^2 + \alpha_1^2\alpha_2 + \alpha_2^3)$$

and $$R(\Delta) = \Lambda\alpha_1/(1-\alpha_2-\alpha_1^2 + \alpha_2^2 + \alpha_1^2\alpha_2 + \alpha_2^3) \tag{2.85}$$

Equation (2.77) then gives

$$R(k\Delta) = \alpha_1 R[(k-1)\Delta] + \alpha_2 R[(k-2)\Delta]$$

where $$k \geqslant 2$$

which produces the complete auto-correlation function.

One point we should note about auto-regressive processes is the possibility of instability due to feedback. Thus, referring to equation (2.82) if, say, $|\alpha| \geq 1$, then R(0) would be negative, and property (2.63) breaks down. This means that we no longer have a stationary random process. If we actually generate such a process we could find that x(t) increases exponentially or oscillates with increasing amplitude. Thus there are restrictions on the values α can take in order to have a stationary random process. For equation (2.85) it can be shown that the condition

$$-1 < \alpha_2 < 1 - \alpha_1, 1 + \alpha_1 \tag{2.86}$$

ensures stability. This is equivalent to saying that the pair of values (α_1, α_2) must be inside a triangular region if plotted on a two-dimensional graph (see Fig. 2.9). Values like $\alpha_1 = \alpha_2 = 1/2$ or $\alpha_2 = 1$ (regardless of what α_1 is) are not permissible, whereas $\alpha_1 = 1$, $\alpha_2 = -1/2$ would ensure stability. Figure 2.10 shows R(kΔ) for this pair of values. It is clear that it satisfies the requirements R(0)>0 and R(0)$\geq$| R(kΔ)| .

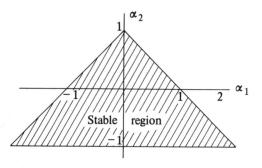

Fig. 2.9   Stability region of second-order auto-regressive processes

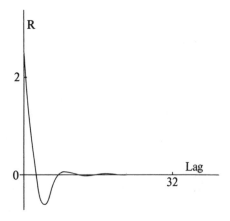

Fig. 2.10   Example of auto-correlation function

We have discussed so far the computation of R given α. In practice finding m and α from measured values of R is the usual case. For this, we combine properties (2.76) with (2.77) for $k = 1, 2, \ldots, m-1$, giving m equations in m unknowns α, and solve them. However, usually we do not know how large m is to start with. An acceptable practice is to start with $m = 1$, find α_1, see how good the fit is, and then go on to $m = 2, 3, \ldots$, until the α's no longer change significantly. For details see [5], Chapter 3. Mathematical discussions can be found in [6] [7] [8].

2.5.4 A CAUTIONARY EXAMPLE*

Let us study another random process, $z(t)$, which

1. has a finite ensemble z_i, $i = 1, 2, \ldots M$.
2. has a constant probability, say Pdt, of making a jump out of its present value into one of the other $M-1$ permissible levels.
3. has a probability, $1-(M-1)Pdt$, that it will not make a jump so that it retains its present value.

Thus, knowing that its value at t is z_i

$$P_z^{t+dt,t}(z_i/z_i) = 1 - (M-1)Pdt$$

$$P_z^{t+dt,t}(z_k/z_i) = Pdt$$

We wish to find $P_z^{t+s,t}(z_i/z_i)$. We derive $P_z^{t+s+ds,t}(z_i/z_i)$ from it as follows: If $x(t+s) = z_i$, then there is chance $1-(M-1)Pds$ that it will remain at z_i. If $z(t+s) = z_i$ there is a chance, Pds, that it will jump into level z_i before $t+s+ds$. Knowing that $z(t) = z_i$, there is chance $P_z^{t+s,t}(z_i/z_i)$ that $z(t+s) = z_i$, and chance $1-P_z^{t+s,t}(z_i/z_i)$ that it is not. Thus we have

$$P_z^{t+s+ds,t}(z_i/z_i) = P_z^{t+s,t}(z_i/z_i) [1-(M-1)Pds] [1-P_z^{t+s,t}(z_i/z_i)]Pds$$

This gives us a differential equation for $P_z^{t+s,t}(z_i/z_i)$, now written simply as $P(s)$ for simplicity

$$\frac{dP}{ds} = \frac{P(s+ds)-P(s)}{ds} = -MPP(s) + P$$

Solving the equation, imposing the initial condition $P_z^{t,t}(z_i/z_i) = 1$, we get

$$P_z^{t+s,t}(z_i/z_i) = \frac{1}{M} - \frac{M-1}{M} \exp[-PMs]$$

We are now able to find the auto-correlation function of z. It is clear from the property of z that all the M levels are equally likely *a priori*. Thus,

$$P_z^t(z_i) = 1/M$$

for every i, which means

$$P_z^{t+s,t}(z_i,z_i) = P_z^{t+s,t}(z_i/z_i)P_z^t(z_i) = M^{-2}[1-(M-1)\exp(-PMs)]$$

whereas

$$P_z^{t+s,t}(z_{k\neq1},z_i) = (1-P(s))(M-1)^{-1}P_z^t(z_i) = M^{-2}[1-\exp(-PMs)]$$

Thus

$$<z(t+s)z(t)> = \sum_{i,k} z_iz_kP_z^{t+s,t}(z_k,z_i)$$

$$= \sum_i (z_i^2/M)\ \frac{1}{M} + \frac{M-1}{M}\exp(-PMs) + \sum_{i\neq k} (z_iz_k/M^2)\ [1-\exp(-PMs)]$$

$$= <z^2>[\frac{1}{M} + \frac{M-1}{M}\exp(-PMs)] + \sum_{i,k}(z_iz_k/M^2) - \sum_1 (z_i^2/M^2)\ [1-\exp(PMs)]$$

$$= <z^2>\exp(-PMs) + <z>^2\ [1-\exp(-PMs)]$$

Suppose z is a zero mean process, then the auto-correlation function is just $<z^2>\exp(-PMs)$.

Recall that the first-order auto-regressive process has

$$R(k\Delta) = \alpha^k\Lambda/(1-\alpha^2) = \exp[-A.k\Delta]\ N/(1-\alpha^2)$$

where $A = -\ln\alpha$, which is a positive number because α is restricted to be less than 1. In other words, the random process we just studied has the same auto-correlation function as an auto-regressive process. Yet the two processes are very different in structure. This example should caution us that, when one is mathematically modelling a physical signal and manages to accurately reproduce the observed auto-correlation function from the model, one should not immediately conclude that the model is a good one. For many practical purposes, however, the auto-correlation is the only quantity available for analysis, but is usually also the only physically important quantity.

2.6 Analysis Methods

2.6.1 THE ERGODIC THEOREM

As we saw in the previous section, we can define the average quantities related to a random process, such as μ,σ, or R, from the theoretical model of the process, either p_x^t and $p_x^{t,s}$ or m and α for auto-regressive processes, or $w(\tau)$ in the case of moving average processes. But what are we to do if we do not have a model? If, say, we are given measured values of x(t) and are required to find a model for it? We saw in Section 2.5.1 how we can find $p_x^{t,s}$

from R for a Gaussian process. In Section 2.5.3 we mentioned the possibility of finding the coefficients of auto-regression from R. The problem is, then, how we can find R and related averages from measured values of x(t).

The avenue for this is opened for us by the **ergodic theorem**, which states that, given the appropriate conditions we can obtain ensemble averages by computing **time averages**. Thus, we have, formally

$$<x(t)> = \overline{x(t)} = \lim_{T\to\infty} \int_{-T}^{T} x(t)dt \qquad (2.87)$$

In other words, it is asserted that, if we observe a random process long enough, we would see that the values that actually occur have the same numerical distribution as p_x. While this might seem so reasonable as to be incontrovertible, there are many cases when the theorem is not valid. For example, if we flip a coin a million times, we are **not** certain to get half a million heads and half a million tails. We do not necessarily get approximately half a million, in fact. It is possible, though very very unlikely, to see a million heads and no tails.

What then are the conditions for the ergodic theorem to be valid? First, x(t) must be a stationary process. As we see, the right-hand side in equation (2.87) is not a function of t at all, and it can reproduce $<x(t)>$ for us only if $<x(t)>$ does not vary with time. But stationarity alone is still insufficient. For example, suppose we have a coin with the strange property that, once it turns a head, it will always turn a head, and same for a tail, though we do not know at the very start whether it will turn head or tail. Thus, *a priori* we have no idea which is more likely, or, $P_x(0) = P_x(1) = 1/2$. Also, $P_x(0/0) = 1$ $P_x(1/1) = 1$ while $P_x(0/1) = P_x(1/0) = 0$. The random process is certainly time-invariant: $P_x(\xi,\eta) = P_x(\xi/\eta)P_x(\eta)$ does not vary with t or s. But since the coin turns heads all the time once it turns a head, and similarly for tail, any x(t) that actually occur would either be a string of 0's or a string of 1's. The time average is either 0 or 1, never 1/2, the ensemble average computed from P_x. Clearly, the ergodic theorem breaks down here.

It is thus clear that, to be ergodic a random process must have an ensemble that contains no **closed domains**, defined as sets of values with $p_x(\xi/\eta)$ that is 0 for any ξ outside a set if η is inside, i.e., once x takes a value inside the set, its chance of taking a future value outside the set is 0. Together, stationarity and absence of closed domains fulfil the essential conditions for ergodicity. In practice, however, the distinction between the two conditions is not that clear. If a random process is stationary but has closed domains in its ensemble, then its properties before and after its entry into a closed domain would be substantially different, and it would appear to be non-stationary.

To make things easy for ourselves, we usually take ergodicity for granted unless there are reasons pointing otherwise, for example, if the source producing the random process violates the two conditions, or if the observed values appear to change their properties with time. Even in these cases we often try to impose ergodicity by making compensatory changes to the values of x(t), as operations like trend removal are designed to do. In short, we take time averages and assume that they are equal to ensemble averages, because we have virtually no other choice!

2.6.2 ESTIMATION OF AVERAGES

The process of computing averages from measured values of a random process is generally called **estimation**. This word carries the implication that we are merely guessing at what the averages are, that the computed results are not guaranteed to equal the theoretical averages. As we saw in the last subsection, the ergodic theorem provides that time averages approach ensemble averages as the observation time interval becomes infinitely long. Since we can never achieve this in practice, the estimates we compute must depart from the theoretical values. In fact, because they are computed from values that fluctuate randomly, they are themselves random variables. They would fluctuate about their theoretical values according to some probability distributions. Our job in estimation is then to devise estimates which are relatively **stable**, meaning that they do not fluctuate too much, and come near their theoretical values most of the time.

Let us consider how to estimate the mean of a stationary random process, $\mu = \ <x(t)>$. First let us say we have measured x(t) once only. Would this value, say x(s), be a good estimate for μ? Clearly, on average x(s) is μ. But this means very little by itself. All this guarantees is that x(s)$-\mu$ is as likely to be positive as negative, or that x(s) is as likely to be too big as too small. On the other hand, if x(s)$-\mu$ has a small mean-square value, then we know that x(s) does not deviate too much from μ most of the time and it would be a good estimate. In other words, we can measure the quality of an estimate by the mean-square deviation of the estimate from its theoretical value.

Given an average value μ and its estimate $\hat{\mu}$, we define the **bias** of the estimate as

$$\beta = \ <\hat{\mu}>-\mu$$

If $\beta = 0$, $\hat{\mu}$ is called an **unbiased** estimate. We have

$$<(\hat{\mu}-\mu)^2> = \ <[(\hat{\mu}-<\hat{\mu}>) - (\mu-<\hat{\mu}>)]^2> = \ <[(\hat{\mu}-<\hat{\mu}>)-\beta]^2>$$

$$= \ <(\hat{\mu}-<\hat{\mu}>)^2> + \beta^2 - 2\beta<\hat{\mu}-<\hat{\mu}>>$$

The first term is just the variance of $\hat{\mu}$, while the last term is always 0. Thus, we can measure the quality of $\hat{\mu}$ as the estimate of μ by looking at the quantity, $d(\hat{\mu})$, with

$$V(\hat{\mu}) + \beta^2 = d(\hat{\mu}) \qquad (2.88)$$

Since both the variance of $\hat{\mu}$ and its bias contribute to the mean-square deviation, in choosing a good $\hat{\mu}$ we often have to trade-off between the two. If, say, we can decrease one substantially at the cost of slightly increasing the other, then the overall quality of the estimate would be improved.

It often happens that the unbiased estimate is intuitively the obvious one, but in fact not the best because of the possibility of reducing the variance by accepting a slight bias. For example, let us examine the possibility of estimating $\mu = <x(t)>$ by $\hat{\mu} = \alpha x(s)$, $0 \leq \alpha \leq 1$. We have

$$\beta = <\hat{\mu}> - \mu = (\alpha - 1)\mu$$

and

$$V(\hat{\mu}) = \alpha^2 V(x) = \alpha^2 \sigma^2$$

Thus

$$d(\hat{\mu}) = \alpha^2 \sigma^2 + (\alpha - 1)^2 \mu^2$$

This is minimised if we take $\alpha = \mu^2/(\mu^2 + \sigma^2)$. The optimal α equals 1 only if σ is 0. To get a clearer picture of the reason for this, suppose that the unbiased estimate falls in the range $\mu - e$ to $\mu + e$, where e is a small number. Obviously, the biased estimate falls in $\alpha(\mu - e)$ to $\alpha(\mu + e)$, (Fig. 2.11). However, the overall deviation is now reduced to $2\alpha e$. This occurs because multiplication by α decreases the upper limit more than it decreases the lower limit. We hasten to add that the above theoretical result is useless in practice because we do not know μ and σ to start with, and would have no idea what value of α is to be used.

Now suppose we have n measured values of $x(t)$, $x_1, x_2, \ldots, x_n$. How do we estimate μ and σ from these? It is usual to estimate μ by the **measured mean**

$$\hat{\mu} = \frac{1}{n} \sum_{i=1}^{n} x_i \qquad (2.89)$$

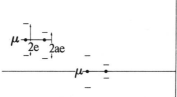

Fig. 2.11 Illustration of reduced error of a biased estimate

and $V(x) = \sigma^2$ by the measured variance

$$\hat{\sigma}^2 = \frac{1}{n-1} \sum_{i-1}^{n} (x_i - \hat{\mu})^2 \tag{2.90}$$

Both are unbiased estimates. (It might seem strange that the dividing factor in definition (2.90) is $n-1$ rather than n. We shall explain this later.) It is, however, not at all easy to find their variances unless some assumptions about their properties are made. From (2.19) we have

$$V(\hat{\mu}) = \sum_{i,k=1}^{n} <x_i x_k>/n^2 - \mu^2$$

which depends on the correlation between different measured values. Assuming for the moment that they are uncorrelated, or $<x_i x_k> = <x_i><x_k>$ if $i \neq k$ then

$$V(\hat{\mu}) = n^{-2} \sum_{i} <x_i^2> + n^{-2} \sum_{i \neq k} <x_i><x_k> - \mu^2$$

Recall that $\quad\quad\quad <x^2> = \mu^2 + \sigma^2, \quad\quad$ thus

$$V(\hat{\mu}) = n^{-2}[n(\mu^2+\sigma^2)] + n^{-2}[n(n-1)\mu^2] - \mu^2 = \mu^2/n \tag{2.91}$$

Since β is 0, this is the same as $d(\hat{\mu})$, which says that $\hat{\mu}$ deviates from μ by $\sigma/\sqrt{n}$ on average, if the n measured values are uncorrelated, or better, independent. It is clear that, by including a large number of uncorrelated measurements we have obtained a good estimate, since it would have a small deviation from its theoretical value.

For a random process, uncorrelated measurements mean that

$$C(|\,t_i - t_k\,|) = 0 \text{ if } i \neq k$$

This is rather difficult to satisfy for all possible indices. Consequently, the actual estimates we have to compute are seldom as good as the above analysis indicates. Suppose we use the following n values of x(t)

$$x(i\Delta), \; i = 0,1,\ldots,n-1$$

Then we have

$$V(\hat{\mu}) = n^{-2} \sum_{i,k=0}^{n-1} R(|\,i-k\,|\,\Delta) - \mu^2$$

$$= n^{-2} \sum_{i,k=0}^{n-1} C(|\,i-k\,|\,\Delta)$$

Put $i-k = m$ we now have

$$V(\hat{\mu}) = n^{-2} \sum_{i=0}^{n-1} \sum_{m=i-n+1}^{i} C(|m|\Delta) \tag{2.92}$$

Let us invert the order of summation, remembering the restrictions $-n+1 \leqslant i-n+1 \leqslant m \leqslant n-1$, which are equivalent to $-n+1 \leqslant m \leqslant n-1$ and $m \leqslant i \leqslant n-1$. The formula then becomes

$$V(\hat{\mu}) = n^{-2} \sum_{m=1-n}^{n-1} C(|m|\Delta) \sum_{i=m}^{n-1} 1 = \frac{1}{n} \sum_{m=1-n}^{n-1} (1-|\frac{m}{n}|)C(|m|\Delta)$$

If, instead of a sum, we wish to estimate μ by the integral

$$\hat{\mu} = \frac{1}{T} \int_0^T x(t)dt$$

Then we can derive an expression similar to equation (2.92) with integration taking the place of summation. Without going into details, we have

$$V(\hat{\mu}) = \frac{1}{T} \int_{-T}^T (1-|\frac{t}{T}|)C(t)dt \tag{2.93}$$

Now let us turn to σ^2. We have

$$<\sigma^2> = \frac{1}{n-1} \sum_{i=1}^n <x_i^2 + \hat{\mu}^2 - 2x_i\hat{\mu}> = \frac{n}{n-1}(<x^2>+<\hat{\mu}^2>-2<\hat{\mu}^2>)$$

By definition, $V(\hat{\mu}) = <\hat{\mu}^2>-\mu^2$, and hence $<\mu^2> = V(\hat{\mu})+\mu^2$. If the x's are uncorrelated then $V(\hat{\mu}) = \sigma^2/n$, and $<\hat{\mu}^2> = \mu^2+\sigma^2/n$. Also by definition, $\sigma^2 = <x^2>-\mu^2$, so that $<x^2> = \mu^2+\sigma^2$. Putting these into the above expression we get

$$<\hat{\sigma}^2> = \frac{n}{n-1}\{(\mu^2+\sigma^2)-(\mu^2+\sigma^2/n)\} = \sigma^2$$

We see that the formula (2.90), with its divisor $n-1$, ensures that the estimate is unbiased. This compensates for the correlation between x and $\hat{\mu}$. But if $V(\hat{\mu})$ is as given in equation (2.92) then $\hat{\sigma}^2$ is no longer unbiased. The exact extent of this, however, is not easy to determine since it depends on C(t).

We can also determine the quality of $\hat{\sigma}^2$ by evaluating

$$d(\hat{\sigma}^2) = <(\hat{\sigma}^2-\sigma^2)^2>$$

that is, its mean-square deviation from what it is meant to estimate. This expression is rather hard to evaluate for a general random process, but manageable for a Gaussian process. We shall state without proof that it simplifies to $2\sigma^4/n$. Again, the larger the sample size n, the better is the quality of the estimate as the mean-square deviation is now smaller.

2.6.3 CONFIDENCE INTERVALS*

As we saw, while the computation of statistical averages from measured values is not difficult, we have the problem of deciding as to how reliable the computed values are. The confidence interval is a way of expressing the reliability of estimates. It is much used in practice, particularly in spectral analysis, often incorrectly.

Consider a Gaussian variable with mean μ and standard deviation σ. The latter is known while the former is to be estimated. We know that

$$\int_{\mu-2\sigma}^{\mu+2\sigma} p_x(\xi)d\xi = 0\cdot95$$

So if we observe one value of x as $\hat{\mu}$, there is 95 per cent chance that it falls within 2σ of μ. But it is just as correct to say that μ must be within 2σ of $\hat{\mu}$ with 95 per cent chance of being correct. Thus, the interval $\{\hat{\mu}-2\sigma, \hat{\mu}+2\sigma\}$ is called the 95 per cent confidence interval of μ.

Now suppose we take n uncorrelated measured values and obtain

$$\hat{\mu} = \frac{1}{n} \sum_{i=1}^{n} x_i$$

We know that $\hat{\mu}$ is a Gaussian random variable with mean μ and standard deviation $\sigma/n^{1/2}$. We thus have $\hat{\mu}-2\sigma/n^{1/2} \leqslant \mu \leqslant \hat{\mu}+2\sigma/n^{1/2}$ at 95 per cent confidence level. Thus, by averaging over a larger set of uncorrelated measurements, we have reduced the confidence interval.

The problem is, however, that usually we do not know σ. We can of course estimate $\hat{\sigma}$ from the measured values of x, and use the result in place of σ. But then the confidence interval itself becomes a random quantity! In this particular case the problem is easily solved. Statisticians have long worked out that the quantity $\hat{\mu}/\hat{\sigma}$ obeys the student-t distribution mentioned in Section 2.4.3. From this distribution we can easily compute a value A such that $\mu/\hat{\sigma}$ falls between $\hat{\mu}/\hat{\sigma}-2A$ and $\hat{\mu}/\hat{\sigma}+2A$ at 95 per cent confidence, which then produces a confidence interval for μ expressed in $\hat{\mu}$, $\hat{\sigma}$ and A. Other situations are unfortunately harder to handle.

A frequently encountered case in spectral analysis is as follows: we have random variable τ which is know to be proportional to a χ^2 variable with n degrees of freedom, i.e., $\tau = A\chi^2_n$. Then, as shown earlier, we know $<\chi^2_n>$ = n and $V(\chi^2_n) = 2n$, we know that τ has mean nA and standard deviation

$A(2n)^{1/2}$. Having taken one measurement of τ, we then have an estimate of its mean. We can then find the constant A using the formula $A = <\tau>/n$. Substituting the value into $A(2n)^{1/2}$, we also have an estimate of the standard deviation of σ, which can then be used to specify the confidence interval. Thus, for large n the χ^2 distribution approaches the Gaussian distribution. Given one value of τ, the confidence interval for its mean is simply $\tau(1\pm(8/n)^{1/2})$. It should be pointed out, however, the formula is useless unless n is well over ten, since otherwise there is considerable difference between the χ^2 and the Gaussian distributions, and in any case the estimate of the standard deviation is unreliable.

References

1. PARZEN, E. *Modern Probability Theory and its Applications*. John Wiley, New York, 1960.
2. FELLER, W. *An Introduction to the Theory of Probability and its Application*. John Wiley, New York, 1968.
3. BENDAT, J. S., and PIERSOL, A.G. *Random Data: Analysis and Measurement Procedures*. John Wiley, New York, 1971.
4. THOMAS, J. B. *An Introduction to Statistical Communication Theory*. John Wiley, New York, 1969.
5. BOX, G. E., and JENKINS, G. M. *Time Series Analysis: Forecasting and Control*. Holden-Day, San Francisco, 1969.
6. BLOOMFIELD, P. *Fourier Analysis of Time Series: An Introduction*. John Wiley, New York, 1976.
7. HANNAN, E. J. *Multiple Time Series*. John Wiley, New York, 1970.
8. KOOPMANS, L. H. *The Spectral Analysis of Time Series*. Academic Press, New York, 1974.

ADDITIONAL REFERENCES

MILLER, I., and FREUND, J. E. *Probability and Statistics for Engineers*. Prentice-Hall, 1965.
DAVENPORT, W. B., and ROOD, W. I. *An Introduction to the Theory of Random Signals and Noise*. McGraw-Hill, New York, 1958.
BENDAT, J. S. Interpretation and application of statistical analysis for random physical phenomena. *IRE Trans. (Bio-med. Elec.)*, **BME-9**, 31−43, Jan. 1962.
COX, D. R., and LEWIS, P. A. W. *The Statistical Analysis of Series of Events*. Methuen, London, 1966.

Chapter 3

Series and Transformation

3.1 Introduction

In this chapter we will consider the analysis of functions in the frequency domain where the information is presented as a discrete series. The main tool for such analysis is the **Fourier transform**.

The analytical methods introduced by Fourier occupy a significant place in the solution of problems in physics and engineering. Originally derived for the analysis of problems in heat engineering, the properties of Fourier analysis are now widely applied in all branches of physics. These properties provide a method for determining the frequency components of a time-varying function. For example, a Fourier analysis of a signal derived from a time-varying displacement can provide the amplitudes of the frequencies present in the original process. The lowest frequency in this type of analysis is known as the fundamental frequency and the higher frequencies, which are integer multiples of the fundamental frequency, are known as the higher-order harmonics. Where the function is periodic, a Fourier series is used to obtain amplitudes of the fundamental and harmonic frequencies present. For this reason Fourier series analysis is often termed **harmonic analysis**.

Fourier analysis can also be extended to non-periodic functions. Here the **period** of analysis (but not the signal being analysed) is assumed to be infinitely long. This gives rise to the use of Fourier integral transformation where all frequencies are presented instead of just those which are integer products of the fundamental frequency.

These types of analyses can also be used relating to functions which are not time-varying. For example, the variation of thickness of rolled steel sheet as a function of one or more of its spatial dimensions can be analysed to determine the characteristic periodicities present and so assist in location of the source of the thickness irregularities.

Finally, the methods of Fourier analysis can be applied to the analysis of discrete data series which results in a form of finite Fourier analysis which has important applications in digital computer calculations.

3.2 Fourier Series Expansion

Fourier's theorem states that a function, $x(t)$ may be expanded over the interval $[0,T]$ as a series of the form

$$x(t) = C_0 + C_1 \sin(\omega_0 t + \theta_1) + C_2 \sin(2\omega_0 t + \theta_2)$$
$$+,\ldots,+ C_n \sin(n\omega_0 t + \theta_n) \tag{3.1}$$

where $\omega_0 = \dfrac{2\pi}{T}$

which may be considered as the sum of the fundamental frequency component, or first harmonic, $C_1 \sin(\omega_0 t + \theta_1)$, having a fundamental frequency of $f = \omega_0 / 2\pi$, together with its harmonics and a constant term C_0. The coefficients $C_1,\ldots, C_n$ represent the peak amplitude excursion of the fundamental and harmonic components of the series. The angles $\theta_1,\ldots,\theta_n$ represent the phase relationship between the initial vector value for the fundamental and those of the harmonics at time t.

A feature of such a data series is that the value of the function, $x(t)$, *exactly* repeats itself at regular intervals, such that

$$x(t) = x(t \pm nT) \tag{3.2}$$
$$n = 0, 1, 2, 3,\ldots$$

The interval required for one complete variation is known as the period T and the number of cycles per unit time is the **fundamental frequency**

$$f_0 = \frac{\omega_0}{2\pi} = \frac{1}{T}$$

This is a fundamental definition, and is frequently needed. A more compact form for the general series given by equation (3.1) is obtained by substituting the relationship

$$\sin(A + B) = \sin A \cdot \cos B + \cos A \cdot \sin B$$

so that we can write

$$x(t) = C_0 + C_1 \sin \theta_1 \cdot \cos \omega_0 t + C_1 \cos \theta_1 \cdot \sin \omega_0 t$$
$$+,\ldots, + C_n \sin \theta_n \cdot \cos n\omega_0 t + C_n \cos \theta_n \cdot \sin n\omega_0 t \tag{3.4}$$

To further streamline the notation, we write $C_n \sin \theta_n$ as a_k and $C_n \cos \theta_n$ as b_k, and by letting $C_0 = a_0/2$, we have

$$x(t) = a_0/2 + a_1 \cos \omega_0 t + b_1 \sin \omega_0 t$$
$$+,\ldots, + a_n \cos n\omega_0 t + b_n \sin n\omega_0 t \tag{3.5}$$

or

$$x(t) = \frac{a_0}{2} + \sum_{k=1}^{n} a_k \cos k\omega_0 t + \sum_{k=1}^{n} b_k \sin k\omega_0 t \qquad (3.6)$$

Equation (3.6) is known as the **Fourier series** for the function x(t).

The reason for making this change is to permit the instantaneous amplitudes for a_0, a_k and b_k to be derived theoretically by the use of certain integrals. It will be obvious that the area under a sinusoidal or cosinusoidal wave-form for a complete period $T = 1/f_0$ must be zero, i.e.,

$$\frac{1}{T} \int_0^T \sin n\omega_0 t \, . \, dt = -\frac{1}{n\omega_0 T} \left| \cos n\omega_0 t \right|_0^T$$

$$= -\frac{1}{n\omega_0 T} [\cos 2\pi n - \cos 0] = 0 \qquad (3.7)$$

and similarly for

$$\frac{1}{T} \int_0^T \cos n\omega_0 t \, . \, dt$$

where n is an integer.
Also writing

$$\cos m\omega_0 t \, . \, \cos n\omega_0 t = \tfrac{1}{2}(\cos(m+n)\omega_0 t + \cos(m-n)\omega_0 t) \qquad (3.8)$$

we see that, if m and n are unequal integers, then

$$\frac{1}{T} \int_0^T \cos m\omega_0 t \, . \, \cos n\omega_0 t \, . \, dt = 0 \qquad (3.9)$$

Similarly

$$\frac{1}{T} \int_0^T \sin m\omega_0 t \, . \, \sin n\omega_0 t \, . \, dt = 0 \qquad (3.10)$$

$$\frac{1}{T} \int_0^T \sin m\omega_0 t \, . \, \cos n\omega_0 t \, . \, dt = 0 \qquad (3.11)$$

The only integrals of this type which have a finite value are

$$\frac{1}{T} \int_0^T \sin^2 n\omega_0 t \, . \, dt = \frac{1}{T} \int_0^T \cos^2 n\omega_0 t \, . \, dt = \tfrac{1}{2} \qquad (3.12)$$

Equations (3.9) to (3.12) define a peculiar property of sinusoids known as

orthogonality, and can be summarised by stating that a finite value is obtained for the weighted average of the product of sine and sine or cosine and cosine if their frequencies and phase shifts are identical, and a zero result if they are not. Also, the weighted average of the product of sine and cosine will be zero irrespective of their frequencies.

This result is important and is the key to many analysis techniques used in signal processing.

It therefore follows from equation (3.5) that

$$\int_0^T x(t).dt = \int_0^T \frac{a_0}{2}.dt = \frac{a_0}{2}.T$$

since the integrals of all terms, other than $a_0/2$, will be zero (equation 3.7). Therefore, the constant term

$$\frac{a_0}{2} = \frac{1}{T} \int_0^T x(t).dt \qquad (3.13)$$

Also multiplying $x(t)$ by $\sin k\omega_0 t$, we have

$$\int_0^T x(t)\sin k\omega_0 t.d_t = \int_0^T b_k \sin k\omega_0 t.d_t = b_k T/2$$

since all terms other than the $\sin^2$ term will be zero (equations 3.9 to 3.12). Therefore

$$b_k = \frac{2}{T} \int_0^T x(t)\sin k\omega_0 t.dt \qquad (3.14)$$

Similarly

$$a_k = \frac{2}{T} \int_0^T x(t) \cos k\omega_0 t.dt \qquad (3.15)$$

These terms a_k and b_k, represent the amplitudes of the harmonics present in the original function, $x(t)$, and are known as the **Fourier coefficients**.

3.3   Fourier Spectrum

The values of the Fourier coefficients a_0, a_k, b_k, provide a further means of defining the function $x(t)$ as a series

$$x(t) = F(a_0, a_1, a_2, \ldots, a_n) + F(b_1, b_2, b_3, \ldots, b_n) \qquad (3.16)$$

A plot of these coefficients, representing the magnitude of the harmonic components of x(t), gives information about the frequency content of the function and is known as the **Fourier spectrum**. This is shown in Fig. 3.1 as two separate relationships for a_k and b_k plotted against the common value of coefficients $\pm k$. It should be noted that whereas the cosine coefficients, a_k, exhibit an even symmetry about $k = 0$, the sine coefficients, b_k, show an odd symmetry.

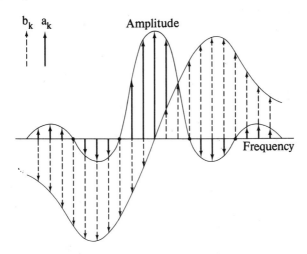

Fig. 3.1   Fourier spectral coefficients

Since the a_k, b_k coefficients represent the same frequency, $k\omega_0$, we can combine them into a single vector sum to represent the modulus and phase of each of the harmonic amplitudes, viz.

$$C_k = \sqrt{(a_k^2 + b_k^2)} \qquad (3.17)$$

and

$$\theta_k = \arctan\left[\frac{b_k}{a_k}\right] \qquad (3.18)$$

We could say that C_k shows the amount of frequency, k, present in x(t) and θ_k is the 'orientation' of its two components.

3.3.1   FOURIER ANALYSIS OF A RECTANGULAR WAVEFORM

As an example of the derivation of the Fourier series and Fourier spectrum, let us consider a single repetition of the rectangular wave-form shown in Fig. 3.2. For reasons given later, this must be made repeatable over a period $0 \rightarrow T$. The sine coefficients are given by equation (3.14).

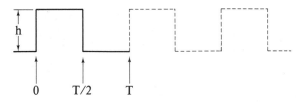

Fig. 3.2   A rectangular wave-form

We may note that x(t) is zero outside the range 0<t<T/2. Also since x(t) has a constant amplitude, h, within this range, we can write

$$b_k = \frac{2}{T} \int_0^{T/2} h \cdot \sin k\omega_0 t \cdot dt$$

Therefore

$$b_k = \frac{2h}{T} \left[\frac{-\cos k\omega_0 t}{k\omega_0} \right]_0^{T/2}$$

and since

$$\omega_0 = 2\pi/T$$

$$b_k = \frac{h}{\pi k} [1 - \cos k\pi]$$

Thus, for unit amplitude of the rectangular wave-form, (h = 1)

$$b_1 = 2/\pi; \ b_2 = 0; \ b_3 = 2/3\pi; \ b_4 = 0; \ b_5 = 2/5\pi; \ \text{etc.}$$

The cosine coefficients, a_k, can readily be shown to be zero from equation (3.15), thus

$$a_k = \frac{2h}{T} \int_0^{T/2} \cos k\omega_0 t \cdot dt = \frac{2h}{T} \left[\frac{\sin k\omega_0 t}{k\omega_0} \right]_0^{T/2}$$

which is zero for all integer values of k.

The mean value coefficient $a_0/2$ is obtained from equation (3.13) as

$$\frac{a_0}{2} = \frac{1}{T} \int_0^{T/2} h \cdot d_t = \frac{h}{2}$$

Hence the Fourier series for the rectangular wave-form, shown in Fig. 3.2, is given as

$$x(t) = \frac{h}{2} + \frac{2h}{\pi} \sin \omega_0 t + \frac{2h}{3\pi} \sin 3\omega_0 t + \frac{2h}{5\pi} \sin 5\omega_0 t + \ldots, \quad \frac{2h}{n\pi} \sin n\omega_0 t$$

(3.19)

The result of plotting this function for the first five terms is shown in Fig. 3.3. This indicates that $x(t)$ is a periodic function of time with a basic periodicity of T seconds equivalent to a fundamental repetition rate of $f = 1/T$ per second.

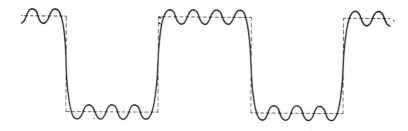

Fig. 3.3 Synthesis of the rectangular wave-form

Thus, the Fourier series expansion over the integration interval T is identical to the expansion of the continuous function $x(t)$ taken over any period of time. The process of Fourier series transformation, therefore, implicitly assumes that the transformed time function repeats itself indefinitely outside the period, T, of the observed function, $x(t)$, at a rate equal to the fundamental frequency $f_0 = 1/T$. In other words, continuous Fourier analysis can only be carried out if the function is periodic or assumed to be periodic.

3.4 Complex Representation

Complex representation of Fourier series and the Fourier integral simplifies the notation and allows later developments of the Fourier transform.
Referring to equation (3.6) we can expand the term

$$(a_k \cos k\omega_0 t + b_k \sin k\omega_0 t)$$

by the use of

$$\cos k\omega_0 t = \tfrac{1}{2}[\exp(jk\omega_0 t) + \exp(-jk\omega_0 t)] \qquad (3.20)$$

and

$$j \sin k\omega_0 t = \tfrac{1}{2}[\exp(jk\omega_0 t) - \exp(-jk\omega_0 t)] \qquad (3.21)$$

where $j = \sqrt{-1}$
thus

$$(a_k \cos k\omega_0 t + b_k \sin k\omega_0 t) = \frac{a_k}{2} [\exp(jk\omega_0 t) + \exp(-jk\omega_0 t)]$$

$$+ \frac{b_k}{2} [\exp(jk\omega_0 t) - \exp(-jk\omega_0 t)]$$

$$= A_k \cdot \exp(jk\omega_0 t) + B_k \exp(-jk\omega_0 t) \qquad (3.22)$$

where

$$A_k = \frac{a_k - jb_k}{2} \text{ and } B_k = \frac{a_k + jb_k}{2} \qquad (3.23)$$

A_k and B_k represent the **complex conjugate amplitude coefficients** of the Fourier series and will now be expanded by the use of equations (3.14) and (3.15) to find the complex Fourier series. For the purposes of clarity, the variable of integration in these equations will be changed to p since we will need to retain t for time in the exponential relationship. This is permissible since the coefficients a_k and b_k are numbers represented by definite integrals. Therefore

$$A_k = \frac{1}{T} \int_0^T x(p)[\cos k\omega_0 p - j \sin k\omega_0 p] \, dp$$

$$= \frac{1}{T} \int_0^T x(p)\exp(-jk\omega_0 p) \, dp \qquad (3.24)$$

and similarly

$$B_k = \frac{1}{T} \int_0^T x(p)\exp(jk\omega_0 p) \, dp \qquad (3.25)$$

Substituting equations (3.22), (3.24), (3.25) and (3.13) in equation (3.6) we obtain an expression for the complex Fourier series

$$x(t) = \frac{1}{T} \int_0^T x(p) \, dp + \sum_{k=1}^{n} \left[\frac{1}{T} \int_0^T x(p)\exp(-jk\omega_0 p) \cdot dp \right] \exp(jk\omega_0 t)$$

$$+ \sum_{k=1}^{n} \left[\frac{1}{T} \int_0^T x(p)\exp(jk\omega_0 p) \cdot dp \right] \exp(-jk\omega_0 t) \qquad (3.26)$$

A simplification is possible since the expression represents a succession of terms; $\exp(jk\omega_0 p) \cdot \exp(-jk\omega_0 t)$ and $\exp(-jk\omega_0 p) \cdot \exp(jk\omega_0 t)$ both extending from k = 1 to k = n. The sign of k can be reversed in the second set of terms by summing from k = −1 to k = −n so that the joint summation can be represented as a succession of terms: $\exp(jk\omega_0 p) \cdot \exp(-jk\omega_0 t)$ with k extending from $k = -n$ to $k = +n$. This now includes the term $k = 0$, when the summed terms reduce to

$$\frac{1}{T} \int_0^T x(p) . dp$$

which is equal to the constant term in equation (3.26) and therefore can be omitted from the resultant expression. Hence equation (3.26) can be written as

$$x(t) = \sum_{k=-n}^{k=n} \frac{1}{T} \left[\int_0^T x(p)\exp(-jk\omega_0 p) . dp \right] \exp(jk\omega_0 t) \quad (3.27)$$

or

$$x'(t) = \int_{k=-n}^{k=n} A_k \exp(jk\omega_0 t) \quad (3.28)$$

This represents a complex Fourier series for the time-history, $x(t)$ limited to $2n+1$ terms.

3.4.1 GIBBS PHENOMENON

Let us now consider the truncated Fourier series $x'(t)$ given by equation (3.28) in terms of the original function, $x(t)$.

We can regard equation (3.28) as representing the summation of the products of two series, A_k and an exponential series which we will call $v(t)$ where

$$v(t) = \sum_{k=-n}^{k=n} \exp(jk\omega_0 t) = \sum_{k=-n}^{k=n} [\exp(j\omega_0 t)]^k \quad (3.29)$$

This is a geometric series of $2n+1$ terms, with leading term, $\{\exp(j\omega_0 t)\}^n$ and each succeeding term being $\exp(-j\omega_0 t)$ times the previous term. Such a series is readily summed by the expression

$$S_m = a_0 + a_1 + \ldots + a_m = a_0(1 - f^{m+1})/(1 - f) \quad (3.30)$$

where

$$f = a_{k+1}/a_k = \exp(-j\omega_0 t)$$

and m is the number of terms minus one, i.e. $m = 2n$. Thus we have

$$v(t) = \exp(j\omega_0 n \frac{t}{T}) \left\{ 1 - \exp[-j\omega_0(2n+1)\frac{t}{T}] \right\} / [1 - \exp(-j\omega_0 \frac{t}{T})]$$

$$= \{\exp[j\frac{\omega_0}{2}(2n+1)\frac{t}{T}] - \exp[-j\frac{\omega_0}{2}(2n+1)\frac{t}{T}]\}/[\exp(j\frac{\omega_0 t}{2T}) - \exp(-j\frac{\omega_0 t}{2T})]$$

$$= \sin[(2n+1)\pi\frac{t}{T}]/\sin(\pi\frac{t}{T}) \qquad (3.31)$$

since

$$\exp(j\omega_0 t) = \cos(\omega_0 t) + j.\sin(\omega_0 t) \qquad (3.32)$$

The function given by equation (3.31) takes the form shown in Fig. 3.4. Note that we have shown its periodic extension outside the interval (0,T). Clearly, $v(t) = 0$ at $t = \frac{PT}{2n+1}$, $p = 1,2,\ldots,2n$, but its value at 0 and 1 is $2n+1$. As t moves towards $t = \frac{T}{2}$ we get a series of peaks of alternating signs and decreasing height. The peak centred at $t = 0$ has a base width of $2T/(2n+1)$, while other peaks have half that width.

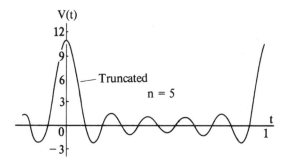

Fig. 3.4   A truncated Fourier series

Further

$$\int_0^T v(t)dt = \Sigma\int \exp(j\omega_0 kt)dt = T \qquad (3.33)$$

since only the $k = 0$ term gives a non-zero contribution.
   It can be shown that $x'(t)$ is related to $x(t)$ by

$$x'(t) = \int_0^T v(t')x(t - t')dt'$$

or, it is a weighted integral of x.
   If $x(t)$ is continuous and varies slowly, then most of the alternating peaks tend to cancel, so that $x'(t)$ is approximately that of $x(t)$ averaged over an area covered by the main peak. Since $x(t)$ varies slowly, its average over a

short interval does not differ appreciably from the exact values, so that x'(t) would approximate x(t) well. However if x(t) contains a discontinuity, then its integral with v(t) would oscillate rapidly in the neighbourhood of the discontinuity. Figure 3.5 shows such a function and its Fourier series approximation. Not only is the error large near the discontinuity, we have also the fact that error does not decrease if we take a larger n. The peaks in the error curve get narrower as n increases, but remain at about the same values. This is why mathematicians say 'Fourier series do not converge uniformly, only in the mean-square sense', which means that the error does not go to zero everywhere as the series gets larger, but if you square the error and integrate over the whole interval, the result does go to zero. (This is because the error curve peaks get narrower all the time.) The failure of Fourier series to converge at discontinuities is given the name **Gibbs phenomenon**.

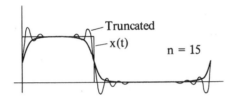

Fig. 3.5   Fourier series approximation of a discontinuous function

In practical computation using finite series we always obtain a truncated series, x'(t) rather than the infinite series, x(t) and the Gibbs phenomenon will be present to a greater or lesser degree. Its effect can be minimised by using 'windowing' techniques which are discussed in the next chapter.

3.5   The Fourier Integral Transform

The Fourier series transformation, described earlier as a means of analysis in the frequency domain, has two major limitations which prevent its application to many time-series of practical interest. In the first case, it assumes that the time function is of **infinite duration**, whereas practical data are often a transient having finite duration; and secondly, the assumption is made that the data are **periodic** over an unlimited extent, whereas practical data are usually non-periodic, as well as being limited in duration.

It is possible in many cases to represent non-periodic data by means of a **Fourier integral transform** which will be defined in this section. This will be referred to as a **Fourier transform** and with its use we may perform harmonic analysis, and obtain power spectra, correlation functions, filtered data and coherence functions. The Fourier transform plays a very important part in the analysis of vibration, shock and other random data.

Intuitively we realise that it will be necessary to extend the range of integration to infinity in order to include the complete transient time-history, since we can no longer assume that the data will repeat themselves outside the range of integration, previously limited to one fundamental period. Thus, we let $k, T \rightarrow \infty$ so that the fundamental frequency will tend towards zero, i.e. $f \rightarrow \delta f \rightarrow 0$. The Fourier coefficients will become continuous functions of frequency as the separation between the harmonics tends to zero and $k\omega_0 = \frac{2\pi k}{T} \rightarrow \omega$. ω is a finite number, as both k and $T \rightarrow \infty$. $\frac{k}{T}$, now a continuous number, is the frequency f. We note that f and ω are related as

$$\omega = 2\pi f$$

This is another fundamental definition we shall require frequently. Thus considering the variable in terms of frequency, rather than complex frequency, we can restate equation (3.27) as

$$x(t) = \int_{-\infty}^{\infty} df \left[\int_{-\infty}^{\infty} x(p)\exp(-j\omega p) . dp \right] \exp(j\omega t)$$

(3.34)

$$= \int_{-\infty}^{\infty} \left[\int_{-\infty}^{\infty} x(p)\exp(-j\omega p) . dp \right] \exp(j\omega t) . df$$

which is a complex form of the Fourier integral.

The term in brackets represents the amplitudes corresponding to the complex Fourier coefficients of a continuous time-series. This is a frequency function and we can write

$$X(f) = \int_{-\infty}^{\infty} x(p)\exp(-j\omega p) . dp$$

The variables in this definite integral can now be changed back to their original time form, viz.

$$X(f) = \int_{-\infty}^{\infty} x(t)\exp(-j\omega t) . dt$$

(3.35)

to give the **complex Fourier transform**. also known as the **complex spectrum** or, simply **Fourier transform** of the time series, x(t). The absolute value of this function yields the frequency amplitude content and the argument gives the phase content.

Equation (3.34) now becomes

$$x(t) = \int_{-\infty}^{\infty} X(f)\exp(j\omega t).df \qquad (3.36)$$

which is known as the **inverse complex Fourier transform**, or **inverse transform** of the frequency series, $X(f)$ and in discrete form

$$x(t) = \sum_{-\infty}^{\infty} X_k(f)\exp(jk\omega_0 t) \qquad (3.37)$$

These forms of the Fourier transform find wide acceptance for practical applications [1] but for analytical purposes it is convenient to use the variable ω in place of f when it is necessary to add a scaling factor, $1/2\pi$ to the complex transform.

If the function, $x(t)$ is an even function, so that it is symmetric about the t $= 0$ axis, then the Fourier transform becomes

$$X_c(f) = \int_{-\infty}^{\infty} x(t) \cos \omega t . dt \qquad (3.38)$$

which is known as the **Fourier cosine transform**, or cospec (f).

For odd functions

$$X_s(f) = \int_{-\infty}^{\infty} x(t) \sin \omega t . dt \qquad (3.39)$$

which is known as the **Fourier sine transform**, or sinspec (f).

3.6 Discrete Fourier Series

For the analysis of digitised data, it is necessary to consider a finite version of the Fourier series developed earlier, and to derive a discrete form of the Fourier transform. The classical Fourier series described in Section 3.4 is identical computationally to the discrete Fourier transform, although its theoretical derivation is quite different. We will consider its derivation following a consideration of the finite Fourier series.

If we consider the calculation of Fourier series and integrals from a discontinuous discrete data series, a number of limitations will be apparent which are not present when the time-history series is processed in continuous form. Broadly, these limitations stem from the need to preserve information in its conversion from a continuous to a discontinuous series. Let us consider first the effect of these limitations on the derivation of a Fourier series used to express a finite ensemble of discrete data points.

Assuming a sample record of finite length T seconds, where T may be termed the fundamental period of the data, we consider the record to be divided into N equally spaced points, having adjacent points separated by a duration, h seconds. The series for x(t) can be represented as a coefficient series having the form

$$x(t) = dc\ term + sine\ terms + cosine\ terms$$

in which only a limited number of frequency components are represented. This limit is determined by the **Nyquist frequency**, f_N, which is related to the sampling interval $f_N = 1/2h$. This may be understood by first noting that only a limited number of time points can exist at which data are able to be present.

These are

$$t = h.i \quad (i = 1,2,3,\ldots,N)$$

where the total length of the record is $T = Nh$. There are only N independent numbers from which we can compute at most N independent results. Now, each Fourier coefficient contains two parts: real and imaginary. Consequently the frequency components must also be limited to a discrete number ($= N/2$) — a consequence of the Nyquist limitation given above

$$f = f_n \quad [n = 1,2,3,\ldots,(N/2)]$$

A rigorous proof of this limitation may be found in many standard engineering texts. Thus if we replace the fundamental frequency term, ω_0 in equation (3.6), by $2\pi/T$, recognising that only a given number of discrete values of frequency are possible, we can write for the series

$$x_i = \frac{a_0}{2} + \sum_{n=1}^{N/2} a_n.\cos \frac{2\pi nih}{T} + \sum_{n=1}^{N/2} b_n.\sin \frac{2\pi nih}{T} \qquad (3.40)$$

or, since $h/T = 1/N$ and the mean value is a special case of summation when $n = 0$, we can also write for x_i

$$x_i = \sum_{n=0}^{N/2} a_n.\cos \frac{2\pi in}{N} + \sum_{n=0}^{N/2} b_n.\sin \frac{2\pi in}{N} \qquad (3.41)$$

which is a definition for the discrete Fourier series.

The coefficients a_0, a_n and b_n are obtained in a similar manner to the derivation of equations (3.13), (3.14) and (3.15), as

$$\frac{a_0}{2} = \frac{1}{N} \sum_{i=1}^{N} x_i \qquad (3.42)$$

$$a_n = \frac{2}{N} \sum_{i=1}^{N} x_i \cos \frac{2\pi in}{N} \tag{3.43}$$

$$b_n = \frac{2}{N} \sum_{i=1}^{N} x_i \sin \frac{2\pi in}{N} \tag{3.44}$$

where $n = 1, 2, \ldots, N/2$.

3.7 The Discrete Fourier Transform

In order to derive a form of Fourier transform for discrete data from the continuous form given in equations (3.35) and (3.36) it is necessary to

(a) replace integrals by sums,
(b) recognise that the limits of summation cannot be infinite.

We saw in the previous section that these limitations result in a discrete form of Fourier series representation which passes through all the sampled data values in the real discrete time-history, x_i. We can see that in order to use these equations we must let the spectral series be complex, thus

$$X_n = (a_n + jb_n) \tag{3.45}$$

which would represent *both* positive and negative frequencies.
Consequently equation (3.41) can be written as the complex Fourier series

$$x_i = \sum_{n=-N/2}^{N/2} X_n \exp(j2\pi in/N) \quad (i = 1, 2, 3, \ldots, N) \tag{3.46}$$

where x_i and X_n are periodic functions. However, since the spectrum is given in complex form there are two spectral components generated for each real frequency. Consequently the summation of the pairs of components in the Fourier transform will result in a doubling of the amplitude of the spectral series produced. It will be seen later that this contravenes **Parseval's theorem** which states that the mean power of the signal must be equal to the sum of the powers contributed by each spectral component. The Fourier transform must therefore include a scaling factor of $1/N$ as shown below

$$X_n = \frac{1}{N} \sum_{i=-N/2}^{N/2} x_i \exp(-j2\pi in/N) \tag{3.47}$$

Equations (3.46) and (3.47) form a Fourier transform pair suitable for expressing a discrete data series.

Their equivalence can be proved by substituting (3.47) into (3.46) and applying the orthogonal relationship described in Section 3.6 which, for complex frequencies, is given as

$$x_i = \sum_{n=-N/2}^{N/2} \exp(j2\pi in/N) \quad \begin{array}{l} = N \text{ if } n = 0, \pm N, \pm 2N \text{ etc.} \\ = 0 \qquad \text{otherwise.} \end{array} \quad (3.48)$$

It may be noted that the scaling factor, $1/N$, appears in the literature in either transform and is sometimes given as $1/\sqrt{N}$ in both transforms. Since the equations constitute a transform pair, these variants are equally valid providing consistency is maintained through their manipulation.

A simplified form of these equations can be obtained by noting that the series x_i and X_n are symmetrical for positive and negative values of N and include the zero value so that we can write

$$X_n = \frac{1}{N} \sum_{i=0}^{N-1} x_i \exp(-j2\pi in/N) \qquad (3.49)$$

and

$$x_i = \sum_{n=0}^{N-1} X_n \exp(j2\pi in/N) \quad [i,n = 0,1,\ldots,(N-1)] \qquad (3.50)$$

These equations will be referred to as the **discrete Fourier transform** (DFT) and the **inverse discrete Fourier transform** (IDFT) respectively. The latter is, of course, the discrete Fourier series itself. A consequence of this notation when implemented on the digital computer using methods described below is that a sequential calculation of spectral values of X_n by a digital computer will be obtained as shown in Fig. 3.6(a) compared with the more familiar representation of the continuous spectrum given in Fig. 3.6(b).

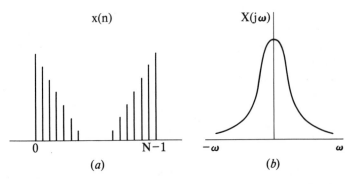

(a)   (b)

Fig. 3.6   Fourier spectrum and discrete Fourier spectrum

A major use of the discrete Fourier transform is the translation of a time series into an equivalent frequency series. We saw earlier than an N-point series in the time domain will transform into a N/2-point series in the frequency domain. It must not be expected however that the useful frequency range will be uniformly distributed over these N/2 points and this can lead to an apparent loss of definition unless a much larger number of time samples are taken. An example is that of a Gaussian time function (N = 800) which we would expect to retain its function shape following transformation into the frequency domain. As shown in Fig. 3.7 this will be expressed rather poorly by about six samples. A solution to this problem is interpolation by addition of zeros to the time-history before transformation takes place.

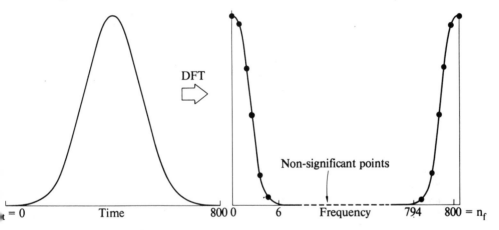

Fig. 3.7 Illustrating data reduction during transformation from time to frequency domain

3.8 The Fast Fourier Transform

In the previous section expressions for Fourier transform pairs were described which are used for the calculation of a discrete time or frequency series. The remainder of this chapter will be concerned with practical methods of calculation using the digital computer.

The discrete complex Fourier transform is a 1:1 conversion of any sequence x_i, $i = 0,1,2,\ldots,N-1$ consisting of N complex numbers, into another sequence defined by

$$X_n = \sum_{i=0}^{N-1} x_i W^{in} \tag{3.51}$$

where $W = \exp(-2\pi j/N)$

Digital Methods for Signal Analysis

The factor $1/N$ will be omitted in this and later derivations in order to simplify notation. It serves as a scaling factor to the values of X_n obtained and needs to be considered only when numerical values are to be attributed to the frequency (or other) scale.

The direct calculation of equation (3.51) requires the solution of a matrix of terms, $[X_n] = [W^{in}].[x_i]$ which are all complex. For example if there are eight samples in the time series $(N = 8)$, then the matrix takes the form

$$
\begin{vmatrix} X_0 \\ X_1 \\ X_2 \\ X_3 \\ X_4 \\ X_5 \\ X_6 \\ X_7 \end{vmatrix}
=
\begin{vmatrix}
W^0 & W^0 & W^0 & W^0 & W^0 & W^0 & W^0 & W^0 \\
W^0 & W^1 & W^2 & W^3 & W^4 & W^5 & W^6 & W^7 \\
W^0 & W^2 & W^4 & W^6 & W^8 & W^{10} & W^{12} & W^{14} \\
W^0 & W^3 & W^6 & W^9 & W^{12} & W^{15} & W^{18} & W^{21} \\
W^0 & W^4 & W^8 & W^{12} & W^{16} & W^{20} & W^{24} & W^{28} \\
W^0 & W^5 & W^{10} & W^{15} & W^{20} & W^{25} & W^{30} & W^{35} \\
W^0 & W^6 & W^{12} & W^{18} & W^{24} & W^{30} & W^{36} & W^{42} \\
W^0 & W^7 & W^{14} & W^{21} & W^{28} & W^{35} & W^{42} & W^{49}
\end{vmatrix}
\times
\begin{vmatrix} x_0 \\ x_1 \\ x_2 \\ x_3 \\ x_4 \\ x_5 \\ x_6 \\ x_7 \end{vmatrix}
$$

If matrix multiplication is carried out in this way we would have to calculate 64 complex products and 64 complex additions. In the general case N^2 complex multiplications and additions would be required. For large values of N this results in extremely lengthy calculation times on the digital computer.

An alternative method of obtaining the discrete Fourier transform, termed the fast Fourier transform (FFT) method, was described first in a form suitable for machine calculation by Cooley and Tukey in 1965 [1]. This is a recursive algorithm, based on the factorisation of the above matrix into a series of 'sparse' matrices (i.e. simpler matrices having many zero terms). Using this technique much of the redundancy in the calculation of repeated products, required by equation (3.51), can be removed to enable a large reduction in calculation time to be realised.

This method is a rediscovery of the earlier work by Danielson and Lanczos [2] which found little application using the desk calculator methods of the time, and had to await the advent of the large-scale digital computer for its successful implementation.

Several other fast methods have since been published [3] using essentially similar ideas to the original algorithm. They are all called fast Fourier transforms and will be distinguished from each other by the originators' initials (e.g. C−T algorithm). The key to these methods lies in their exploitation of the possibilities for factorising the number of values of the series to be transformed. An explanation will be given first in terms of matrix factorisation and followed by the derivation of a general computing algorithm. Finally a third description will be given providing a working algorithm suitable for high-level language implementation.

Commencing with the matrix description of the method we note that series representation of a variable is considered generally in terms of a long series of coefficients. We can equally well describe the series in terms of r rows and s columns of a matrix, in which the total number of points is $N = rs$.

A Fourier transform of the elements of this matrix into an equivalent set of Fourier coefficients can then be arranged by carrying out s parallel Fourier transforms, each of r data points, on the individual columns and then summing the results. We have thus reduced the problem to a summation of a number of much shorter Fourier transforms instead of the calculation of one long transform. Quite apart from the saving in time (which will be shown below), this method has another major advantage in that a much smaller storage space is required to hold the intermediate calculations.

To illustrate the working of the algorithm in matrix terms a simple two-level factorising for the total number of terms, $N = rs$, will be considered first. Using this assumption we can express the N variables n and i in equation (3.51) as four subseries for i_0, i_1, n_0, and n_1

$$n = n_1 r + n_0 \qquad (3.52)$$

and

$$i = i_1 s + i_0 \qquad (3.53)$$

where

$$
\begin{aligned}
n_0 &= 0,1,2,\ldots,r-1 \\
n_1 &= 0,1,2,\ldots,s-1 \\
i_0 &= 0,1,2,\ldots,s-1 \\
i_1 &= 0,1,2,\ldots,r-1
\end{aligned}
$$

We can see that i and n each still contain N discrete values if we take some particular values and substitute these in the above. E.g. if $N = 20 = rs$, where $r = 5$ and $s = 4$, then

$$
\begin{aligned}
&n_0 \quad \text{goes from 0 to 4} \\
&n_1 \quad \text{goes from 0 to 3} \\
&i_0 \quad \text{goes from 0 to 3} \\
&i_1 \quad \text{goes from 0 to 4}
\end{aligned}
$$

and a table for n will include all 20 values as required. This is shown in Table 3.1

Table 3.1

$n_1 =$	0	1	2	3	n_0
$n =$	0	5	10	15	0
	1	6	11	16	1
	2	7	12	17	2
	3	8	13	18	3
	4	9	14	19	4
			$r = 5$		

derived from equation (3.52). Similarly for i from equation (3.53) shown in Table 3.2.

Table 3.2

$i_1 =$	0	1	2	3	4	i_0
$i =$	0	4	8	12	16	0
	1	5	9	13	17	1
	2	6	10	14	18	2
	3	7	11	15	19	3
			$s = 4$			

Equation (3.51) can now be rearranged as two sums

$$X_{(n_1,n_0)} = \sum_{i_0=0}^{i_0=s-1} \sum_{i_1=0}^{i_1=r-1} x_{(i_1,i_0)} W^{in} \qquad (3.54)$$

This is a recursive formula in which we carry out complex multiplications and additions at every s'th sample of the $N - 1$ possible variables for i within the inner loop, and then advance the starting point for the first value of i in the outer loop. The inner loop procedure is then repeated from this new point (see Fig. 3.8).

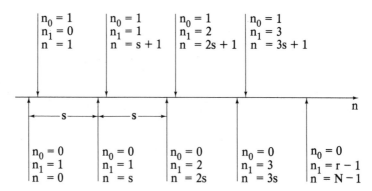

Fig. 3.8 Fast Fourier transform — simple partitioning

If we now expand W^{in} using equations (3.52) and (3.53) we have

$$W^{in} = W^{(i_1 s + i_0)(n_1 r + n_0)}$$
$$= W^{i_1 n_1 rs} . W^{n_0 i_1 s} . W^{i_0(n_0 + n_1 r)} \qquad (3.55)$$

but $i_1 n_1 rs = i_1 n_1 N$ and, since i_1, n_1 is an integer by definition, the exponent is a multiple of 2π and

$$W^{i_1 n_1 N} = \exp \left(-j \frac{2\pi}{N} \right)^{i_1 n_1 N} = 1 \qquad (3.56)$$

We may also note for later use that

$$W^{N/2} = \exp(-j2\pi/2) = -1 \qquad (3.57)$$

Equation (3.54) now becomes

$$X_{(n_1 n_0)} = \sum_{i_0=0}^{i_0=s-1} \sum_{i_1=0}^{i_1=r-1} x_{(i_1,i_0)} W^{n_0 i_1 s} \cdot W^{i_0(n_0+n_1 r)} \qquad (3.58)$$

The inner sum over i_1 depends only on i_0 and n_0 and can be defined as a new series

$$x_{i(i_0,n_0)} = \sum_{i_1=0}^{i_1=r-1} x_{(i_1,i_0)} W^{n_0 i_1 s} \qquad (3.59)$$

which we can recognise as the Fourier transform of a reduced array

$$x_{(i_1,i_0)} \cdot$$

Equation (3.58) can now be rewritten as

$$X_{(n_1,n_0)} = \sum_{i_1=0}^{i_1=s-1} x_{1(i_0,n_0)} W^{i_0(n_0+n_1 r)} \qquad (3.60)$$

There are N elements in array x_1, but, unlike the original array x_i, we only require r operations to evaluate the transform, giving a total of Nr operations for array x_1 (remembering that these are complex operations). Similarly we require Ns operations in order to calculate x_1 from x_i. Consequently, this simple two-step algorithm requires a total of $T = N(r + s)$ complex operations. We can see how this procedure may be extended by increasing the number of steps to p when the number of operations will be $T' = N(s_1 + s_2 + \ldots, s_p)$, where $N = s_1, s_2 \ldots, s_p$. If $s_1 = s_2 = s_3 \ldots, = s_p = s$ then $N = s^p$, so that s can be expressed as the radix of the number series and we can write $p = \log_s N$ from which

$$T' = sN \log_s N = N \left(\frac{s}{\log_2 s} \right) \log_2 N \qquad (3.61)$$

A plot of s against $s/(\log_2 s)$ indicates that radix $s = 3$ is the most efficient, although for practical reasons in digital calculation a radix of two simplifies the algorithm. Using this radix an improvement in computing speed over direct evaluation of

$$\frac{N^2}{2N\log_2 N} = \frac{N}{2\log_2 N} \qquad (3.62)$$

is obtained. For large values of N this saving in computing time can be quite considerable.

The improvements in computational speed obtained by FFT methods have been extended to hardware logic design and a number of special purpose machines have been constructed to implement the FFT in this way.

Experience in the use of these processors indicates that the increase in speed and reduction in running costs, compared with software implementation on a general purpose computer, can be very considerable [4]. The increase in speed enables the hardware FFT processor to be used in a real-time environment such as radar or sonar signal processing and for realisation of real-time digital filtering.

3.8.1 TRANSFORMS WHEN N = 2^p

The practical advantages gained by using a radix of two for the algorithm are also obtained by making the radix equal to a power of two. In the case of $N = 2^p$ intermediate transforms are reduced to two elements and, from equations (3.56) and (3.57), two of the three exponentials shown in the expression of equation (3.55) will assume one of the two values $+1$ and -1, thus avoiding some of the need for complex arithmetic. Similar advantages are obtained for $N = 4^p$ when the exponentials obtained are ± 1 and $\pm j$. The implementation of the FFT as described by Cooley and Tukey required that N be a power of two. This algorithm will be developed below by considering it as a general recursive device programmed by the digital computer. The original series is converted in stages, each stage providing an intermediate transform which supplies the input for subsequent stages, until the final series becomes the discrete transform of the original series.

With N limited to a power of two it is possible to express i and n in terms of the index, p, as a **binary-weighted series**

$$i = i_{p-1}2^{p-1} + i_{p-2}2^{p-2} + ,\ldots, + i_1 2 + i_0 \qquad (3.63)$$

$$n = n_{p-1}2^{p-1} + n_{p-2}2^{p-2} + ,\ldots, + n_1 2 + n_0 \qquad (3.64)$$

In this way we can express all the N possible values of the indices in terms of a binary number and hence facilitate the consideration of storage in the digital computer. The input and output series of the Fourier transform, x_i and X_n are considered as being stored within the memory at location addresses defined by the binary representation of i and n. Using this convention, equation (3.51) can be written in terms of a factorised sum, exactly as carried out previously for $N = rs$, but with the factorising expressed in a binary-weighting form, i.e.

$$X_{(n_{p-1},\, n_{p-2},\ldots,\, n_0)} = \sum_{i_0=0}^{i_0=1} \sum_{i_1=0}^{i_1=1} , \ldots, \sum_{i_{p-1}=0}^{i_{p-1}=1}$$

$$x_{(i_{p-1},\, i_{p-2},\ldots,\, i_0)} \cdot W^{n(i_{p-1}\, 2^{p-1} + i_{p-2}\, 2^{p-2} +,\ldots,\, i_0)} \tag{3.65}$$

That this is equivalent to equation (3.51) can be seen if we consider a simple case of

$$N = 2^2 = 4_{10} = \; \diagup 1 \quad \underset{\mid}{0} \quad 0_{\diagdown 2}$$
$$\qquad\qquad\qquad i_2 \qquad i_1 \qquad i_0$$

In the decimal case

$$\sum_{i=0}^{i=4} x_{i_{(10)}} = x_0 + x_1 + x_2 + x_3$$

and in the binary case

$$\sum_{i_0=0}^{1} \sum_{i_1=0_2}^{1} \sum_{i=0}^{1} x_{i_{(2)}} = x_{000} + x_{001} + x_{010} + x_{011} + x_{100}$$

which is seen to contain all the possible values of x_i in the range up to decimal 4, expressed in binary form.

If we now expand the first term of the exponential for equation (3.65) as we did in equation (3.55)

$$W^{n i_{p-1}\, 2^{p-1}} = W^{n_{p-1}\, 2^{p-1}\, i_{p-1}\, 2^{p-1}} \cdot W^{n_{p-2}\, 2^{p-2}\, i_{p-1}\, 2^{p-1}}$$

$$.,\ldots,.\, W^{n_1 2\, i_{p-1}\, 2^{p-1}} \cdot W^{n_0\, i_{p-1}\, 2^{p-1}} \tag{3.66}$$

we find that all the intermediate terms go to unity and

$$W^{n i_{p-1}\, 2^{p-1}} = W^{n_0 i_{p-1}\, 2^{p-1}} \tag{3.67}$$

This enables the innermost sum of equation (3.66) over $i_p - 1$ to be written as a shorter Fourier transform.

$$X_{1(n_0,\, i_{p-2},\ldots,\, i_0)} = \sum_{i_{p-1}=0}^{i_{p-1}=1} x_{(i_{p-1},\, i_{p-2},\ldots,\, i_0)}\, W^{n_0 i_{p-1}\, 2^{p-1}} \tag{3.68}$$

which depends only on $n_0, i_{p-2}, \ldots, i_0$.

Unlike the complete Fourier transform, this sum consists of a set of N numbers only, each calculated from two of the original data points. Subsequent sums, proceeding outwards in equation (3.65), can be calculated using a generalised recursive expression for the exponential term

$$W^{n i_{p-q}\, 2^{p-q}} = W^{(n_{q-1}\, 2^{q-1} + n_{q-2}\, 2^{q-2+},\ldots,\, n_0)\, i_{p-q}\, 2^{p-q}}$$

$$(q = 1,2,3,\ldots,p) \tag{3.69}$$

The successive sums are evaluated according to the equation

$$X_{q(n_0, n_1, \ldots, n_{q-1}, i_{p-q-1}, i_{p-q-2}, \ldots, i_0)}$$

$$= \sum_{i_{p-q}=0}^{i_{p-q}=1} X_{q-1(n_0, n_1, \ldots, n_{q-2}, i_{p-q}, \ldots, i_0)}$$

$$W^{(n_{q-1} 2^{q-1} + n_{q-2} 2^{q-2} +, \ldots, n_0) i_{p-q} 2^{p-q}}$$

$$(q = 1,2,3,\ldots,p) \tag{3.70}$$

which is the definition of the C−T algorithm.

To apply this recursive formula the initial set of data x_i is first made equal to x_0 $(q = 1)$, thus

$$x_i = x_{(i_{p-1}, i_{p-2}, \ldots, i_0)} = x_{0 (i_{p-1}, i_{p-2}, \ldots, i_0)}$$

This leads to the derivation of succeeding arrays in x_q so that the final array will be that for X_n. Since we have represented the two sets of p arguments for x_q, namely $(n_0, \ldots, n_{q-1}, i_{p-q-1}, \ldots, i_0)$, as binary representations of their storage locations, this can be used to simplify the working of the algorithm and so reduce the storage requirements for the intermediate sums. This involves fetching values from these two storage locations, carrying out complex multiplication and addition in accordance with equation (3.70) and putting the results back into these same two locations. Over-writing the input array with output values thus occurs. This indexing scheme has one disadvantage in that the elements of the final array in X_n are stored in the incorrect order in the core memory. Thus for the last array calculated

$$X_{(n_{p-1}, n_{p-2}, \ldots, n_0)} = X_{p(n_0, \ldots, n_{p-2}, n_{p-1})} \tag{3.71}$$

This shows that, in order to find X_n, the order of the bits in the binary representation of n must be reversed so as to obtain the index of the memory location where X_n may be found from the x_p array. From equation (3.70) we see that for each element of x_q, one complex multiply/add operation is required. Also, a further complex multiply/add operation is required for the recursive generation of the complex exponential. Thus, the total number of operations is $T' = 2Np$. But, since there are $N = 2^p$ elements in each array x_q where $q = 1, 2, 3, \ldots, p$, the improvement in speed over direct evaluation is $T/T' = N/(2.\log_2 N)$, which is the result obtained earlier for the general case.

An alternative form of the algorithm is obtained if the roles of the indices i and n are interchanged when carrying out the expansion of the exponential term in equation (3.66). This is known as the Sande−Tukey version [5] and leads to the following recursion

$$X'_{q(n_0, n_1, \ldots, n_{q-1}, i_{p-q-1}, i_{p-q-2}, \ldots, i_0)}$$

$$= \sum_{i_{p-q}=0}^{i_{p-q}=1} X_{q-1(n_0, n_1, \ldots, n_{q-2}, i_{p-q}, \ldots, i_0)}$$

$$\cdot W^{(i_{p-q} 2^{p-q} + \ldots, i_0) n_{q-1} 2^{q-1}} \tag{3.72}$$

which is the definition of the S–T algorithm.

The two algorithms are operationally very similar although some advantages in speed, particularly for in-place array manipulation, are claimed for the S–T algorithm, due to its simpler exponential structure.

3.8.2 PROGRAMMING THE ALGORITHM

Although the recursive algorithms described previously can be implemented directly, the alternative derivation, given below, is preferred since this will avoid the unnecessary calculation of some complex exponentials. This derivation also assumes that N is a power of two.

Referring to the definition of the discrete Fourier transform given by equation (3.51) if we consider the series x_i to be divided into two interleaved series

$$x_{(2i)} = Y_i$$

$$x_{(2i+1)} = z_i \quad [i = 0, 1, \ldots, (N/2 - 1)] \tag{3.73}$$

where y_i consists of the even-numbered samples and z_i the odd-numbered samples, the discrete Fourier transforms of these two series can now be written (omitting the scaling factor $2/N$) as

$$Y_n = \sum_{i=0}^{(N/2)-1} y_i W^{2in} \tag{3.74}$$

$$Z_n = \sum_{i=0}^{(N/2)-1} z_i W^{2in} \tag{3.75}$$

Note that i, n now represents a half-length series so that $W_{N/2} = \exp[(-2\pi j)/(N/2)] = W^2_N$. The transformed series Y_n and Z_n are displaced by one sampling interval so that to obtain the discrete Fourier transform from the complete N-point series using equations (3.74) and (3.75) we write

$$X_n = \sum_{i=0}^{(N/2)-1} [y_i W^{n(2i)} + z_i W^{n(2i+1)}]$$

$$= \sum_{i=0}^{(N/2)-1} y_i W^{2in} + W^n . \sum_{i=0}^{(N/2)-1} z_i W^{2in} \tag{3.76}$$

since W^n is a constant for a given value of n. Hence

$$X_n = Y_n + W^n Z_n \tag{3.77}$$

But n is limited to $(N/2) - 1$ samples so that to complete the sequence for X_n we require to find the terms from $N/2$ to $N - 1$. We note from the theory of the discrete Fourier transform that for $n > N/2$ the transforms Y_n and Z_n will repeat periodically the values obtained with $0 < n \leqslant N/2$ so that $(n + N/2)$ can be substituted for n in the phase shift term only, viz.

$$X_{(n+N/2)} = Y_n + W^{n+N/2} Z_n \tag{3.78}$$

But

$$W^{n+N/2} = \exp(-j2\pi n/N - j\pi),$$

which, from equation (3.57)

$$= -\exp(-j2\pi n/N) = -W^n \tag{3.79}$$

Hence

$$X_{(n+N/2)} = Y_n - W^n Z_n \tag{3.80}$$

Equations (3.77) and (3.80) enable the first and last $N/2$ points to be evaluated from the separate transforms formed from $x_{(2i)}$ and $x_{(2i+1)}$. Neglecting addition and subtraction as taking negligible computing time compared with multiplication, we see that this result requires $N+2(N/2)^2$ multiplications compared with N^2 for direct evaluation. For large values of N this represents a reduction by almost half of the computing time required.

The process of dividing the series by two and then obtaining the discrete Fourier transform from the transforms of each half of the series can be repeated to obtain further reductions in computing time. In the limit, providing that the series is capable of division by two, a two-point transform is obtained from two single-point transforms. The discrete Fourier transform of a single point is, however, the point itself, thus

$$X_n = \sum_{i=0}^{i=0} x_i W^{in} = x_i \qquad (i=0)$$

so that the method reduces to a simple series of additions and multiplication of an exponential factor, W^n: In this limiting case the improvement in computing speed is, $N/(2.\log_2 N)$ as obtained previously. This version of the algorithm is called the **successive doubling method**, and, as described above, has been termed **decimation in time**.

A similar technique divides the sequence for x_i directly into two non-interleaved sequences

$$x_i = y_i$$

$$x_{i+N/2} = z_i \quad (i=0, 1,\ldots,(N/2-1))$$

and interleaves the values of the final sequences obtained for X_n. This is called **decimation in frequency**.

A graphical description of the successive doubling method illustrating the formation of a suitable computer algorithm is the **signal flow graph** shown in Fig. 3.9. This consists of a series of nodes or points, each representing a variable as the sum of other variables originating from the left of the diagram and connected together by means of straight lines. These additions may be weighed by a number appearing at the side of an indicating arrow on the connecting lines. Thus, from Fig. 3.9 the variable A7 is derived from variables originating at nodes B3 and C3, with the latter weighed by a variable, W^2 so that we can write

$$A7 = B3 + W^2C3$$

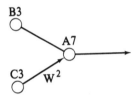

Fig. 3.9   Principle of the signal flow graph

The complete decimation down to pairs of single-value transforms will lead to a shuffling of the data into apparently unrelated pairs of values. This shuffling is due to the repeated regrouping of the position of the intermediate results as indicated in Fig. 3.10 for several small values of number

N = 4		N = 8			N = 16			
0		0	0	0	0	0	0	0
2		1	2	4	1	2	4	8
		2	4	2	3	4	8	4
1		3	6	6	4	6	12	12
3		4	1	1	5	8	2	2
		5	3	5	6	10	6	10
		6	5	3	7	12	10	6
		7	7	7	8	14	14	14
					9	1	1	1
					10	3	5	9
					11	5	9	5
					12	7	13	13
					13	9	3	3
					14	11	7	11
					15	13	11	7
						15	15	15

Fig. 3.10   Shuffling of data

sequences. If the process commences with shuffled data then it becomes possible to carry out computation using a minimum of array storage space. Assuming this to have been carried out (a mechanism for doing this will be suggested later) then the procedure is shown in the single flow graph of Fig. 3.11 for the extraction of the first converted sample, x_0 from a sequence of eight values of x_i. This is seen to produce a summation which, when multiplied by the scaling factor $1/N$ gives

$$X_n = \frac{1}{8} \sum_{i=0}^{i=7} x_i W^{in} = \frac{1}{8} \sum_{i=0}^{i=7} x_i \quad \text{(for } n = 0\text{)}$$

which we know to be correct.

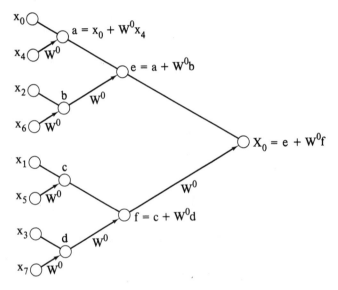

Fig. 3.11   Extraction of the first transform sample

A complete flow graph for an eight-point sequence is given in Fig. 3.12. From this we see that it is not necessary to allocate separate storage for the intermediate arrays of values, a_0, b_0 etc. Storage is only required sufficient to accommodate the initial number of data points, and a similar area for the calculation of product values (since each calculated value is needed twice). The first set of intermediate transforms can be calculated sequentially and placed in the locations previously occupied by the first set of values, which are now no longer required. The second set can then be calculated and returned to the locations occupied by the first set, and so on until the final transform is formed. The procedure is called **in-place** computation. The complex exponentials required can be calculated in advance and stored in a

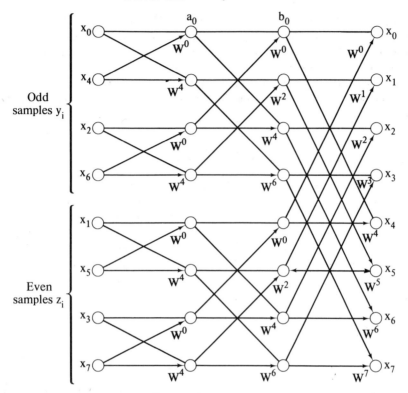

Fig. 3.12   Signal flow graph for eight sampled data values

table, or computed during the calculation of the intermediate transforms. Standard sine/cosine subroutines are generally employed although the use of a recursive method, such as that suggested by Singleton [6], can lead to faster computation.

The data shuffling required in this version of the algorithm corresponds to the bit-reversal noted earlier. If the value of i is written in binary form, then the reversed order of the digits, retranslated back to decimal form, will give the correct order for the location of the input-sample. Note that this initial data shuffling can also be carried out 'in-place' as described previously.

It is possible to rearrange the flow graph to produce a **bit-reversed transform** from correctly-ordered data (as is the case for the original C−T algorithm). In this case the powers of W, needed in the computation are required in bit-reversed order. A version can be obtained in which both input and transformed output data can be in the correct natural order. However, in this case the computation can no longer be carried out 'in-place' and at least double the array storage is necessary [7].

A two-part procedure for a Fortran FFT is shown in Fig. 3.13. A reordering section shuffles the location of the data into a bit-reversed order and a second section carries out the FFT calculation using repeated calls to the exponential subroutine from the main loop.

This program applies to the calculation of the discrete Fourier transform but can be used for the calculation of the inverse transform with trivial alterations.

3.8.3 TRANSFORMATION OF REAL DATA*

The basic FFT algorithms described above have been developed for complex data series. When applied to practical time-history analysis real data will be available and hence transforms of a real data series will be required. We can, of course, set the imaginary parts to zero and insert the relevant data samples in the real part of the synthesised complex sequence. However, this will be wasteful both of computer time and storage space. A more efficient procedure makes use of the conjugate symmetry of a real variable.

If a time-history series x_i is real then

$$x_i = x^*_i$$

and

$$X_n = X^*_{-n} = X(^*N{-}n) \tag{3.81}$$

This latter relationship follows from the periodic nature of the exponent W in equation (3.51)

$$W^{in} = W^{(n+N)i} = W^{(i+N)n} \tag{3.82}$$

Hence

$$x_i = x_{(kN + i)}$$
$$X_n = X_{(kN + n)} \qquad (k = 0, 1, \ldots, \text{etc.})$$

Therefore

$$X_{-1} = X_{(N-i)}$$
$$X_{-n} = X_{(N-n)} \tag{3.83}$$
$$X^*_{-n} = X^*_{(N-n)}$$

This means that if x_i is real then we only need *half* the data points in order to specify the frequency spectrum.

If we consider the real data series to be separated into two interleaved series, as defined in equation (3.73) defining y_i as the real part, and z_i as the imaginary part of a complex function x_i, we can write: $e_i = (y_i + jz_i)$. The discrete Fourier transform of this will give (neglecting scaling factor, $1/N$)

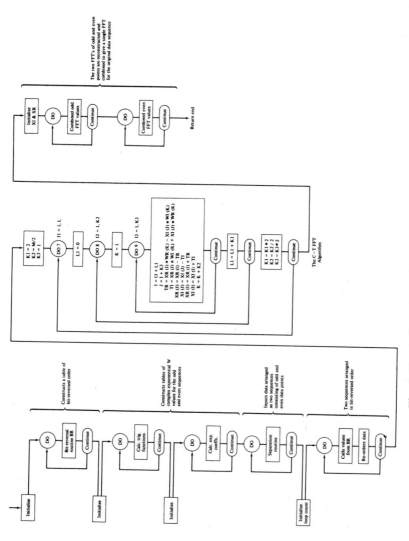

Fig. 3.13   A flow diagram for a fast Fourier transform

$$X_n = \sum_{i=0}^{(N/2)-1} (y_i + jz_i)W^{2in}$$

$$= \sum_{i=0}^{(N/2)-1} y_i W^{2in} + j. \sum_{i=0}^{(N/2)-1} z_i W^{2in}$$

which we can represent as

$$X_n = Y_n + jZ_n$$

But from equation (3.77)

$$X_n = Y_n + W^n Z_n \quad [n = 0,1,\ldots, (1/2N - 1)]$$

Each of these transforms will contain $1/2N$ points, and since the original series y_i and z_i were real, then from equation (3.81) only $1/4N$ points will be required to describe them.

Hence

$$Y_{(1/4N-m)} = Y^* \tag{3.84}$$

and

$$Z_{(1/4N-m)} = Z^*_{(1/4N+m)} \quad [m = 0,1,\ldots, (1/4N - 1)]$$

Also

$$X^*_n = Y_n - jZ_n \tag{3.85}$$

so that by adding and subtracting equation (3.83) and (3.85) we can obtain

$$X_n + X^*_n = 2Y_{(1/4N-m)} \tag{3.86}$$

and

$$X_n - X^*_n = 2jZ_{(1/4N-m)}$$

This enables Y_n and Z_n to be obtained from our original transform in X_n and substituted in equation (3.77) to obtain X_n for the first $1/4N$ points.

The second $1/4N$ points are obtained from

$$X_{(1/2N-m)} = Y_n - W^n Z_n \tag{3.87}$$

A similar procedure can be carried out to evaluate simultaneously the Fourier transforms of two equal length sets of real data. Thus if we let the two series be x_1 having a Fourier transform X_1 and x_2 having a Fourier transform X_2, then substituting these arrays to form the real and imaginary parts of a complex series x_i (note than an interleaved series is not involved here): $x_i = x_1 + jx_2$ and we obtain the transform

$$X_n = X_1 + jX_2 \tag{3.88}$$

From equation (3.81) we can replace n by $N - n$ so that taking the complex conjugate of both sides we obtain

$$X^*_{(N-n)} = X_1 - jX_2 \qquad (3.89)$$

giving

$$\left.\begin{aligned}X_1 &= \tfrac{1}{2}[X_n + X^*_{(N-n)}] \\ X_2 &= \tfrac{1}{2}(X_n - X^*_{(N-n)}) \cdot [n = 0,1,\ldots,(\tfrac{1}{2}N-1)]\end{aligned}\right\} \qquad (3.90)$$

Since X_1 and X_2 are real the symmetry property of equation (3.81) applies and only half of each array need be computed and stored.

Thus only the same amount of storage will be required to evaluate X_1 and X_2 taken together as would be needed to obtain a single complex transform, X_n. Computation time will be only slightly increased by the necessity to form sums and differences using equation (3.90).

3.8.4 COMPUTING THE FFT

Computing an eight-point discrete transform through the use of the fast Fourier transform algorithm takes three steps, each step requiring eight multiplications and eight additions. In general, the operations count for an N-point FFT equal to $N.\log_2 N$, compared with N^2 when equation (3.51) is used directly. As an example of possible savings, if $N \simeq 1000$ then the FFT demands 100 times less effort than the simpler method, and this ratio increases with N. It may also be shown that round-off error can be reduced by a similar factor of $\log_2 N/N$ [8]. There are however a number of disadvantages of the FFT method which are worth examining.

In the first instance it is not economic to use the FFT if only a small number of transformed values are required. Due to the way the algorithm functions it requires only fractionally more effort to produce a complete set of transformed values. Using equation (3.51) directly requires N operations for each transformed value required.

Secondly the FFT cannot be implemented by software in real-time without recourse to specially designed hardware. Using equation (3.51) directly we can bring in one value of x_i into the computer at a time, multiplying by W^{in} for every value of i and adding the $N-1$ products to $N-1$ accumulators. We cannot do this with the FFT. As shown earlier we need to process N input values together to produce N further numbers which are again processed together in the next stage and so on. The requirement to process all N values of x_i at the beginning of the FFT process precludes real-time operation without employing parallel processing hardware.

For the FFT to be possible, N must be a highly composite number. The case when N is a power of two is commonly assumed since this gives almost the largest possible savings in computing time. However, this is sometimes inconvenient since often the length of input is fixed by other factors and may be somewhere between two powers of two. It is then necessary to add a string of zeros to the data in order to make the length a power of two. This is equivalent to abruptly terminating the data, and gives rise to leakage. Consequently it is sometimes desirable to choose an N which contains non-binary factors, e.g. $N = 2^n 3^m$. The FFT algorithm for such values of N is somewhat more complex and also less efficient [9].

The actual calculation of the FFT is more complex than the simpler method. Whereas equation (3.51) can be computed entirely in real arithmetic, complex multiplication is involved in the FFT giving four separate products of real and imaginary values. A further complexity lies in the order of the resulting output vector. In many cases this is in bit-reversed order so that a reordering has to be carried out on the transformed data. Finally the cost of extra programming compilation and de-bugging needs to be taken into account when considering the use of the FFT.

Taking into account all these additional computing costs, we find that the FFT gives no real savings in computing time for $N < 50$ and where N is less than this then the simpler method is to be preferred.

3.9   Applicability of Non-Sinusoidal Series*

Apart from the sinusoidal functions there exist a number of other orthogonal series which enable domain processing to be carried out. The most important of these from our point of view are the **Walsh** and **Haar** function series.

They are characterised by assuming only **two** states, thus matching the behaviour of digital logic, and yet possess many of the attractive manipulative properties of the sine/cosine series.

Historically the Haar series was described first by Alfred Haar in 1909 [10]. He proposed a set of orthogonal functions, taking essentially only two values such that the formal expansion of a given continuous function in the new functions converges uniformly to the given function. This was a property not possessed by any orthogonal set known up to that time and his proposals served to emphasise the unifying theories on orthogonal series developed earlier.

The Walsh functions defined in 1923 by J. L. Walsh [11] form a complete orthogonal set and, although taking only the values $+1$ and -1 were found to have many properties similar to the trigonometric series. At about the same time, but independent of Walsh, another set of two-level orthogonal functions were presented by H. Rademacher [12] which were found later to form an incomplete but true subset to the Walsh functions.

These three function sets form the basis of a new direction in digital processing which have applicability to certain areas of operations such as, for example, the processing of sampled and digitised images.

The functions form inumerably infinite sets of periodic orthogonal square-wave functions which are characterised by having piecewise constant value between innumerably infinite jump discontinuities. We shall, however, only be concerned with finite and discrete sets of such functions.

3.9.1 THE WALSH FUNCTION SERIES

The Walsh functions form an ordered set of rectangular wave-forms taking only two amplitude values, $+1$ and -1, defined over a limited time interval, T, known as the **time-base**. Two arguments are required for complete definition, a time period, t, normalised to the time-base as t/T, and an ordering number, n, related to frequency in a way which is described later. The function is written

$$WAL(n,t) \qquad (3.91)$$

For most purposes a set of such functions is ordered in ascending value of the number of zero crossings found within the time-base. Figure 3.14 shows the first eight of these with the ordering arranged in this way.

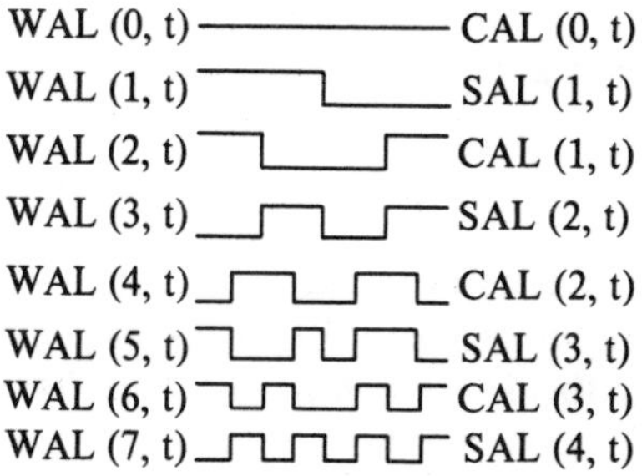

Fig. 3.14   First eight Walsh functions

The function series can be obtained by means of a difference equation, as products of Rademacher functions, by the use of a Hadamard matrix and in other ways [13].

Derivation through the Rademacher function will be described here. A subset of the Walsh functions known as the **Rademacher functions** are obtained from an appropriate set of sinusoidal functions through amplification followed by hard limiting (Fig. 3.15). A complete set of Walsh functions may be obtained from these by taking selected function products in the following way.

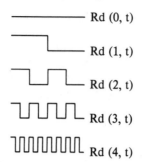

Fig. 3.15   First eight Rademacher functions

If we represent the Rademacher functions as Rd(i,t) where i represents the order by number of zero crossings shown in Fig. 3.15 then a given Walsh function is obtained from

$$WAL(n,t) = \prod_{i=1}^{m} Rd(i,t)^{b_i} \qquad (3.92)$$

Here, n, the order number, is expressed in binary form as a Gray code and b_i gives the relevant bit positions. Hence to find WAL(9,t) we first express nine in binary code as 1001 and then rearrange this in Gray code as 1101. The second bit position gives $b_i = 0$ so that we can write directly from equation (3.92)

$$WAL(9,t) = Rd(4,t) \cdot Rd(3,t) \cdot Rd(1,t)$$

Similarly for other functions in the series.

The orthogonality of the discrete Walsh series can be proved as follows. First an expression for the discrete Walsh function having $N = 2^p$ terms will be stated in terms of a continued product as

$$WAL(n_{p-1}, n_{p-2}, \ldots, n_0; t_{p-1}, t_{p-2}, \ldots, t_0)$$

$$= \prod_{r=0}^{p-1} (-1)^{n_{p-1-r}(t_r + t_{r+1})} \qquad (3.93)$$

where n,t are the arguments of the function expressed in binary notation. The sum of the products of any two discrete Walsh functions is given as the binary summation

$$\sum_{t_{p-1}}^{1} \sum_{t_{p-2}}^{1} \ldots \sum_{t_0=0}^{1} WAL(m_{p-1}, m_{p-2}, \ldots, m_0; t_{p-1}, t_{p-2}, \ldots, t_0)$$

$$\cdot WAL(n_{p-1}, n_{p-2}, \ldots, n_0; t_{p-1}, t_{p-2}, \ldots, t_0) \qquad (3.94)$$

Substituting equations (3.93) in (3.94) gives for the binary product-sum

$$\sum_{t_{p-1}}^{1} \sum_{t_{p-2}}^{1} \cdots \sum_{t_0=0}^{1} \prod_{r=0}^{p-1} (-1)^{(n_{p-1-r}+m_{p-1-r})(t_r+t_{r+1})}$$

$$= \prod_{r=0}^{p-1} \sum_{t_r=0}^{1} (-1)^{(n_{p-1-r}+m_{p-1-r})(t_r+t_{r+1})}$$

$$= \prod_{r\neq 0}^{p-1} [1 + (-1)^{(n_{p-1-r}+m_{p-1-r})}] \qquad (3.95)$$

Now if each $n_t = m_t$, remembering that only two values are possible, 0 or 1, then equation (3.95) becomes

$$\prod_{r=0}^{p-1} (1+1) = 2^p = N \qquad (3.96)$$

If at least one $n_t \neq m_t$ then at least one term in the product given by equation (3.94) is 0 giving a zero product. Hence in terms of decimal indices we have

$$\sum_{t=0}^{N-1} WAL(m,t)WAL(n,t) = \begin{cases} N \text{ for } n = m \\ 0 \text{ for } n \neq m \end{cases} \qquad (3.97)$$

From this the Walsh functions can be seen to form an orthogonal set which can be normalised by division by N to form an orthogonal system.

The ordering shown in Fig. 3.14 is known as **sequency order**. **Sequency** is a term proposed by Harmuth [14] to describe a periodic repetition rate which is independent of wave-form. It is defined as: 'one half of the average number of zero crossings per unit time interval'. From this we see that frequency can be regarded as a special measure of sequency applicable to sinusoidal wave-forms only.

A further notation has also been introduced by Harmuth to classify the Walsh functions in terms of even and odd wave-form symmetry, viz.

$$WAL(2k,t) = CAL(k,t)$$

$$WAL(2k-1,t) = SAL(k,t) \qquad (3.98)$$

$$k = 1,2\ldots(N/2 - 1)$$

defining two further Walsh series having close similarities with cosine and sine series.

Using the Walsh function series any complex wave-form can be represented by superposition of a number of elements of the series having appropriate amplitudes in an exactly analogous manner to Fourier synthesis. It will be found that fewer Walsh functions are required to represent a discontinuous wave-form, compared with synthesis using sine/cosine functions, and that the converse holds true for smoothly-varying wave-forms. This brings out an important difference between the two sets of functions, indicating their respective roles in signal analysis and other applications.

3.9.2 THE WALSH TRANSFORM

Decomposition of a time series function x(t) into a Walsh series is similar to Fourier series representation viz.

$$x(t) = a_0 + a_1 \, WAL(1,t) + a_2 \, WAL(2,t) + \ldots \tag{3.99}$$

where the sequency coefficients are given by

$$a_k = \int_0^1 WAL(k,t)\, x(t)\, dt \tag{3.100}$$

A finite discrete Walsh transform pair can be described for a time series x_i comprised of N samples

$$X_k = \frac{1}{N} \sum_{i=0}^{N-1} x_i \, WAL(k,i) \tag{3.101}$$

$$k = 0,1,2,\ldots,N-1$$

$$x_i = \sum_{k=0}^{N-1} X_k \, WAL(k,i) \tag{3.102}$$

$$i = 0,1,2,\ldots,N-1$$

Similar transforms, $X_C(k)$ and $X_S(k)$ can be obtained using Harmuth's CAL and SAL functions.

The transform and its inverse are obtained by matrix multiplication using the digital computer. Since the matrices are symmetrical for the Walsh transform (unlike the Fourier transform) then both transform and inverse transform are identical, except for a scaling factor, 1/N.

If we compare equation (3.101) with the corresponding discrete Fourier transform

$$X_f = \frac{1}{N} \sum_{i=0}^{N-1} x_i \, \exp(-j2\pi i f/N) \tag{3.103}$$

$$f = 0,1,2,\ldots,N-1$$

we note that whilst WAL(k,i) is real and limited to values ± 1, $\exp(-j2\pi if/N)$ is complex and can assume N different values. As a direct consequence of this the Walsh transform proves considerably easier and faster to calculate using digital methods. A fast Walsh transform algorithm (FWT) exists having comparable computational features to the FFT. If the transform is evaluated directly then N^2 additions/subtractions are required. Using the FWT only $2N\log_2 N$ mathematical operations are required to transform N values, where N is expressed as a power of two. Further the Walsh transform requires only **real** additions and subtractions, compared with the **complex** multiplications needed for the Fourier transform.

The important properties of the Walsh function and the Walsh transform are summarised and compared with the sine/cosine functions in Table 3.3. Important differences, not already considered, concern the effect of circular time shift on the series to be transformed and the special behaviour of transform products.

Table 3.3

	Walsh		Fourier	
time index	i		time index	i
sequency index	k		frequency index	n
		$i,k,n = 0,1,2\ldots\ldots\ldots N-1$		
cal function	$cal(k,i) = wal(2k,i)$		cos function	$\cos \dfrac{2\pi ni}{N}$
sal function	$sal(k,i) = wal(2k-1,i)$		sin function	$\sin \dfrac{2\pi ni}{N}$
Walsh function	$wal(k,i)$		Complex function	$\exp \dfrac{j.\,2\pi ni}{N}$
				$= \cos \dfrac{2\pi ni}{N} + j \sin \dfrac{2\pi ni}{N}$
time series	$x_i = a_0 wal(0,i) + a_1 wal(1,t)$		time series	$x_i = A_0 + A_1 \exp\left[\dfrac{2\pi i}{N}\right]$
	$+\ldots\ldots\ldots+ a_N wal(N,i)$			$+A_2\exp\left[\dfrac{4\pi i}{N}\right]+\ldots+A_N \exp(\pi i)$
sequency coefficients	$a_0, a_1\ldots\ldots\ldots a_N$		complex frequency coefficients	$A_0, A_1\ldots\ldots\ldots A_N$

Table 3.3 (continued)

	Walsh		Fourier	
time index	i		time index	i
sequency index	k		frequency index	n
		$i,k,n = 0,1,2\ldots\ldots\ldots\ldots N-1$		

Walsh series

$$a_i = a_0\, \omega al(0,1) + \sum_{m=1}^{N/2}\sum_{p=1}^{N/2-1} a_m sal(m,i)+b_p cal(p,i)$$

Fourier series

$$a_i = \frac{a_0}{2} + \sum_{m=1}^{N}\left[a_m \cos\left[\frac{2\pi ni}{N}\right] + b_m \sin\left[\frac{2\pi ni}{N}\right]\right]$$

Walsh transform

$$X_k = \frac{1}{N}\sum_{i=0}^{N-1} x_i \omega al(k,i)$$

Fourier transform

$$X_n = \frac{1}{N}\sum_{i=0}^{N-1} x_i \exp\left[-j\frac{2\pi ni}{N}\right]$$

Walsh inverse transform

$$x_i = \sum_{i=0}^{N-1} X_k \,\omega al(k,i)$$

Fourier inverse transform

$$x_i = \sum_{i=0}^{N-1} X_n \exp\left[j\frac{2\pi ni}{N}\right]$$

Walsh CAL transform

$$X_c(p) = \frac{1}{N}\sum_{i=0}^{N-1} x_i\, cal(p,i)$$

Fourier cosine transform

$$X_c(p) = \frac{1}{N}\sum_{i=0}^{N-1} x_i \cos\left[\frac{2\pi ni}{N}\right]$$

$$p = 1, 2\ldots\ldots\ldots\frac{N}{2} - 1$$

Walsh SAL transform

$$X_s(q) = \frac{1}{N}\sum_{i=0}^{N-1} x_i\, sal\,(q,i)$$

Fourier sine transform

$$X_s(p) = \frac{1}{N}\sum_{i=0}^{N-1} x_i \sin\left[\frac{2\pi ni}{N}\right]$$

Orthogonal property

$$\sum_{i=0}^{N-1} \omega al(k,i)\,\omega al(l,i)$$

$$= N \quad k = l$$

$$= 0 \quad k \neq l$$

Orthogonal property

$$\sum_{i=0}^{N-1} \exp\left[j\frac{2\pi ni}{N}\right]\exp\left[j\frac{2\pi mi}{N}\right]$$

$$= N \quad n = m$$

$$= 0 \quad n \neq m$$

Symmetrical property

$$\omega al(k,i) = \omega al(i,k)$$

Symmetrical property

Not applicable

The discrete Fourier transform is invariant to the phase of the input signal so that the same spectral decomposition can be obtained independently of the phase or circular time shift of the input signal. This is not the case with the Walsh transform.

The effect is one of variation in shape and location of the dominant spectral characteristics of the signal. It is of less importance where the sum of the square of pairs of transformed coefficients of the same sequency are taken, as with power spectrum derivation, but will account for minor variations between spectra unless time-histories are obtained at the same time or adjusted for zero phase shift.

An addition relationship is found for Walsh functions, namely

$$\text{WAL}(k,t)\text{WAL}(p,t) = \text{WAL}(k \oplus p, t) \qquad (3.104)$$

where $\oplus$ represents Modulo-2 addition.

This corresponds to the set of relationships

$$2\cos kt \cos pt = \cos(k-p)t + \cos(k+p)t$$

$$\ldots \text{etc.}$$

with the important difference that a **shift theorem** for Walsh functions **does not exist** so that whilst the product of two Fourier transforms can be transformed to obtain a convolution of the two original time series, a similar result is *not* obtained with the Walsh transform.

The absence of a Walsh shift theorem successfully prevents fast correlation via the FWT or applications of digital filtering through use of a defined impulse response function in the time domain.

Despite this major difficulty the Walsh function finds many important applications in processing including spectral analysis and digital filtering.

3.9.3 THE HAAR FUNCTION SERIES

The Haar functions also form an orthogonal and orthonormal system of periodic square waves. The amplitude values of these square waves do not have uniform value, as with Walsh wave-forms, but assume a limited set of values, 0, ± 1, $\pm\sqrt{2}$, ± 2, $\pm 2\sqrt{2}$, ± 4 etc. They may be expressed in a similar manner to the Walsh functions as

$$\text{HAR}(n,t) \qquad (3.105)$$

If we consider the time-base to be defined as $0 \leq t \leq 1$ then, following the simplified definition suggested by Kremer [15], we can write

$$
\begin{aligned}
\text{HAR}(0,t) &= & 1 & \quad \text{for} \quad & 0 \leq t \leq 1 \\[4pt]
\text{HAR}(1,t) &= & \begin{cases} 1 & \text{for} \quad 0 \leq t < 1/2 \\ -1 & \text{for} \quad 1/2 \leq t < 1 \end{cases} \\[10pt]
\text{HAR}(2,t) &= & \begin{cases} \sqrt{2} & \text{for} \quad 0 \leq t < 1/4 \\ -\sqrt{2} & \text{for} \quad 1/4 < t < 1/2 \\ 0 & \text{for} \quad 1/2 < t \leq 1 \end{cases} \quad (3.106)
\end{aligned}
$$

$$HAR(3,t) = \begin{cases} 0 & \text{for} & 0 \leq t < \tfrac{1}{2} \\ \sqrt{2} & \text{for} & \tfrac{1}{2} < t < \tfrac{3}{4} \\ -\sqrt{2} & \text{for} & \tfrac{3}{4} < t \leq 1 \end{cases}$$

$$\vdots \qquad \vdots \qquad \vdots$$

$$HAR(2^p + n,t) = \begin{cases} \sqrt{2^p} & \text{for} & n/2^p \leq t < (n+\tfrac{1}{2})/2^p \\ -\sqrt{2^p} & \text{for} & (n+\tfrac{1}{2})/2^p \leq t < (n+1)2^p \\ 0 & & \text{elsewhere} \end{cases}$$

$$p = 1,2,\ldots,\log_2 N \quad n = 0,1,\ldots,2^p-1 \quad N = 2^p$$

This allows a sequential numbering system analogous to that adopted by Walsh for his function series.

The first eight Haar functions are shown in Fig. 3.16. The first two of these are identical to WAL(0,t) and WAL(1,t). The next function HAR(2,t) is simply HAR(1,t) shifted into the left-hand half of the time-base and modified in amplitude to $\pm\sqrt{2}$. The next function HAR(3,t) is identical but shifted into the right-hand half of the time-base. In general all members of the same function subset are obtained by a lateral shift of the first member along the time axis by an amount proportional to its length.

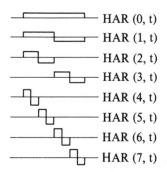

Fig. 3.16   First eight Haar functions

As shown in Fig. 3.16 the essential characteristic of the Haar function is seen as a constant value everywhere except in one sub-interval where a double step occurs.

From the defrnition given in equation (3.106) it can be seen that the Haar functions are also orthogonal, viz.

$$\int_0^1 HAR(m,t)HAR(n,t) = \begin{cases} 1 \text{ for } n = m \\ 0 \text{ for } n \neq m \end{cases} \qquad (3.107)$$

The Haar series can also be shown to be **complete** as is the Walsh series.

There is no theorem analogous to the shift or addition theorems found with Fourier and Walsh series respectively.

3.9.4 THE HAAR TRANSFORM

The discrete Haar transform and its inverse are given by

$$X_n = \frac{1}{N} \sum_{i=0}^{N-1} x_i \cdot HAR(n,i/N) \qquad (3.108)$$

$$x_i = \sum_{n=0}^{N-1} X_n \cdot HAR(n,i/N) \qquad (3.109)$$

$$i,n = 0,1,\ldots,N-1$$

Unlike the Walsh transform the matrix is **not symmetric** so that **separate** transform operations are required for transformation and inverse transformation. If the transform given in equation (3.109) is carried out directly then N^2 additions will become necessary. This can be reduced to $p.N$ where $N = 2^p$ if only the non-zero values are considered.

A considerable improvement in computational efficiency is obtained if a factorisation algorithm similar to that used for the fast Fourier and Walsh transforms is employed.

Using a fast Haar transform (FHT) the total number of additions or subtractions is reduced to

$$N + \frac{N}{2} + \frac{N}{4} + \ldots + 2 = 2(N-1) \qquad (3.110)$$

Transformation time is thus linearly proportional to the number of terms, N, in contrast to Walsh or Fourier fast transformation.

The periodogram definition for the Haar power spectrum is difficult to use with the definition given by equation (3.106) since the equivalent periodicity of a number of adjacent functions can be identical.

If we define the effective sequency of the Haar function series as 'one half the average number of zero crossings per unit time interval' as we did for the Walsh series then it will be seen that the Haar functions fall into discrete groups, each member of a group having the same effective sequency as other members of the same group (see Table 3.4). Using this definition a power spectrum may be defined which can be considered as analogous to the sequency or frequency spectrum by taking the normalised value of the squares of the line spectra that fall within each grouping. The spacing of sequency values will not however be linear although the sparsity of calculated values may be considered acceptable in view of the extremely fast computation that can be achieved [16].

Table 3.4 **Haar sequency groupings**

	Sequency group
HAR(0,t)	1
HAR(1,t)	2
HAR(2,t), HAR(3,t)	3
HAR(4,t) HAR(5,t) ⎫ HAR(6,t), HAR(7,t) ⎬	4
HAR(8,t), HAR(9,t) ⎫ HAR(10,t), HAR(11,t) ⎪ HAR(12,t), HAR(13,t) ⎬ HAR(14,t), HAR(15,t) ⎭	5

References

1. COOLEY, J. W., and TUKEY, J. W. An algorithm for the machine calculation of complex Fourier series. *Math. Comp.*, **19**, 297, 1965.
2. DANIELSON, G. C., and LANCZOS, C. Some improvements in practical Fourier analysis and their application to X-ray scattering from liquids. *J. Franklin, I.*, **233**, 365, 1942.
3. BRIGHAM, E. O. *The Fast Fourier Transform*. Prentice-Hall, 1974.
4. CORINTHIUS, M. J. The design of a class of fast Fourier transform computers. *I.E.E.E. Trans. Computers*, **C-20**, 617−623, June 1971.
5. GENTLEMAN, W. M., and SANDE, G. Fast Fourier transforms for fun and profit. *AFIPS Proc. Fall Joint Comp. Conf.*, **29**, 563, 1966.
6. SINGLETON, R. C. On computing the fast Fourier transform. *Comm. ACM*, **10**, 647, 1967.
7. GLASSMAN, J. A. A generalisation of the fast Fourier transform. *IEEE Trans. (Comp.)*, **C-19**, **2**, 165, 1970.
8. KANEKO, T., and LIN, B. Accumulation of round-off error in fast Fourier transforms. *JACM*, **17**, 637−54, 1970.
9. COOLEY, J. W., LEWIS, A. W., and WELCH, P. D. The fast Fourier transform and its applications. *IEEE Trans. (Ed.)*, **E-12**, **1**, 27, 1969.
10. HAAR, A. Zür Theorie der orthogonalen Funktionensysteme, *Math. Annal.*, **69**, 331−71, 1910.
11. WALSH, J. L. A closed set of orthogonal functions. *Amer. J. Math.*, **45**, 5−24, 1923.
12. RADEMACHER, H. Einige Sätze von allgemeinen orthogonal Funcktionen. *Math. Annal.*, **87**, 112−138, 1922.
13. BEAUCHAMP, K. G. *Walsh Functions and Their Applications*. Academic Press, 1975.
14. HARMUTH, H. F. *Transmission of Information by Orthogonal Functions*. Springer Verlag, 1972.
15. KREMER, H. Algorithms for the Haar functions and the fast Haar transform. Symposium: *Theory and Applications of Walsh Functions*, Hatfield Polytechnic, England, 1971.

16. THOMAS, D. W. Burst detection using the Haar spectrum. Symposium: *Theory and Applications of Walsh Functions*, Hatfield Polytechnic, England, 1973.

ADDITIONAL REFERENCES

CHAMPENEY, D. C. *Fourier Transforms and Their Physical Applications*. Academic Press, 1973.
AHMED, N., and RAO, K. R. *Orthogonal Transforms for Digital Signal Processing*. Springer Verlag, 1975.
HARMUTH, H. F. *Sequency Theory*. Springer Verlag, 1976.

Chapter 4

Spectral Analysis

4.1 Introduction

The statistical techniques developed in Chapter 2 were concerned with the general characteristics of signals. Now that we have discussed the significance of time and frequency series and their interrelation in Chapter 3, we are in a position to consider the statistical characteristics of the signal specifically in the frequency domain.

The results of a frequency analysis can be presented in various ways. Two of the most commonly used quantities are the amplitude of the signal and the power contained in the signal at each frequency, in the form of a frequency spectrum. We shall use the term 'power' in a special sense, as discussed later, even though it is recognised that the data under analysis need not necessarily relate to energy considerations. The term 'auto-spectrum' is sometimes used in place of 'power spectrum' for this very reason. This is purely a question of nomenclature, and mathematically the two are the same.

There are a number of reasons why we wish to compute the power spectrum of a signal. We saw that, owing to the random nature of many physical signals it is necessary to extract from them average quantities which are relatively stable and meaningful to interpret. The power spectrum is an average quantity which tells us about the power or energy content distributed over various frequencies. In cases when the total energy of signals is difficult to measure, but the power at individual frequencies is measurable, the integration of the power spectral density then yields valuable information about the total power. These techniques are used, for example, in shock wave and vibration analysis. Finally, the relation between spectral density and correlation, which is another average quantity related to random data, provides a close link with earlier work in statistical physics and other subjects, permitting us to find analogies of known theories in those areas in signal analysis. In some cases it is possible to measure the power spectrum, and from this the correlation function, even when a signal is itself unmeasurable. (For example, we are not able to measure visible light as a time-series, but we can measure the spectrum of a light source.) This 'indirect route' to the auto-correlation function is particularly important after the fast Fourier transform algorithms became available [1].

At least four basic techniques exist for the derivation of power spectral density:

1. Direct Fourier transform.
2. Indirect method via the Fourier transform of the auto-correlation function.
3. Band-pass filtering.
4. Fourier transform after segmentation.

The direct method involves computing the Fourier transform of the time series and taking the mean-square value of the transform. Until fairly recently the direct method was not applied generally to experimental data because of the high computation cost. However, the development of the FFT has altered this situation. The method has also the honour of being the earliest method used by mathematicians and theoretical physicists.

Calculation of correlation functions, of which we shall have more to say in the next chapter, has been carried out extensively on random data since its introduction by G. I. Taylor in the 1920s. The later work of Wiener [2] and Khintchine [3] showed that the power spectral density and the auto-correlation form a Fourier transform pair and this has provided a method of derivation of the power spectral density by first calculating the auto-correlation and then taking its Fourier transform. A practical approach to this method was developed by Blackman and Tukey [4] and has been used widely in all fields of engineering and physics. This technique proved to be considerably faster to carry out using digital computers than the direct Fourier transform method as the situation stood then. In either case, however, computation of complete spectral density functions was a time-consuming business.

Traditionally filtering methods have been carried out on the analog computer and generally take the form of a set of parallel band-pass filters extending over the frequency range of interest. The signal to be analysed is applied simultaneously to all the filter inputs and the output signals are then squared and averaged to provide discrete points on the power spectral density frequency spectrum. In the case of random signals it is necessary to include division by the filter bandwidth in the averaging process. This is not necessary for periodic signals, since all the energy can be considered to be contained in a series of Dirac peaks. Other analog methods, such as heterodyning the signal with a variable-frequency oscillator using a single narrow-band filter are also in use and considered as equivalent to the filter bank method. Digital filters are further discussed in the next chapter. A method related to filtering is the complex demodulation method mentioned briefly in Chapter 1.

The improvement in cost and speed of digital computing hardware and the more ready availability of signal processing software, and most of all the development of the FFT, have radically changed the situation of relative cost as far as power spectrum computation is concerned. While analog filtering methods continue to be used because of their low cost, and in special applications where for some reason or other digital methods are unusable, current methods of power spectral density evaluation using the direct method of calculation from the FFT of the time-series are more general. In particular, the more recently developed segmentation method appears to offer the best cost effectiveness in terms of time and memory requirement and accuracy of results. However, more recent development in digital hardware [5] indicates that it is now economical to assemble fast parallel arithmetic units in sufficient quantities to permit very rapid calculation. Under these conditions the indirect method may become prominent yet again. Further, as the indirect method produces the auto-correlation function as a by-product, it is extremely useful in preliminary analysis, when we need to extract as much information as possible in order to decide upon the actual computation procedure to be adopted.

4.2 Power Spectrum Estimation

It is common knowledge to physics students that the electric power consumed by a load is proportional to the square of the voltage across it, and that the kinetic energy of a moving mass is proportional to the square of its velocity. The square of a random signal, similarly, describes its 'energy content', and the square of its Fourier transform at frequency f indicates its energy content at that frequency.

However, as we learnt in Chapter 2, a random process has values that fluctuate unpredictably. It follows that the value of its Fourier transform at any frequency would also fluctuate. Thus, if we measure two pieces of the random signal at different times, the Fourier transform of one would differ from that of the other, perhaps considerably so. As a result, the square of the Fourier transform of a *particular* piece of a random signal cannot be used to describe its energy content *in general*. The quantity that does provide a general description is the ensemble average of the Fourier transform square, because it indicates, on average, how important the Fourier transform at any frequency is. Thus, we define the **power spectrum** (or **power spectral density**) of a random process x(t) as

$$S(f) = <| X(f)|^2> \tag{4.1}$$

We recall (Chapter 3) that

$$x(t) = \int_{-\infty}^{\infty} X(f)\exp(j\omega t)df \tag{4.2}$$

(Remember that $\omega = 2\pi f$.) This can be used to derive the **Parseval theorem**, which states that the total power contained in the Fourier transform X(f) equals that in x(t). Take two real functions x(t) and y(t). Their product can be expressed in Fourier transform terms as

$$x(t)y(t) = x(t) \int_{-\infty}^{\infty} Y(f)\exp(j\omega t)df$$

so that

$$\int x(t)y(t)dt = \int\{\int x(t)Y(f)\exp(j\omega t)df\}dt \qquad (4.3)$$

This can be rearranged as

$$\int x(t)y(t)dt = \int Y(f) \{\int x(t)\exp(j\omega t)dt\}df$$

We recognise the bracketed quantity as the complex conjugate of X(f). Hence

$$\int x(t)y(t)dt = \int X^*(f)Y(f)df \qquad (4.4)$$

Now if x(t) = y(t) and X(f) = Y(f), then the above formula becomes

$$\int x^2(t)dt = \int X(f)X^*(f)df = \int |X(f)|^2 df \qquad (4.5)$$

Taking the ensemble average on both sides we get

$$\int S(f)df = \int <x^2(t)>dt \qquad (4.6)$$

Since $<x^2>$ is the mean-square of x it is the average power. Equation (4.6) then shows that the power spectral density added up for all frequencies by integration equals the average power of x added up for all time.

4.2.1 ESTIMATION PROCEDURES

As we said in Section 2.6, in an actual data analysis exercise the required averages are estimated by carrying out some kind of time averaging process. The estimation of S(f) can be carried out in many different ways. We can take a number of time slices of x(t), Fourier transform each, and average over all the slices. This is known as the **segment averaging method**. The alternative general approach is to assume that S(f) varies slowly with f, so that for frequencies $f_1, f_2, \ldots f_n$, $S(f_1) \simeq S(f_2) \simeq \ldots \simeq S(f_n)$, and they can all be obtained by averaging over $|X(f_1)|^2$, $|X(f_2)|^2, \ldots, |X(f_n)|^2$. In other words, after computing $|X(f)|^2$, we perform an averaging operation on it to produce our estimate of S(f). This takes the form of a weighted integration, i.e.

$$S(f) \simeq \int W_f(f')|X(f')|^2 df' \qquad (4.7)$$

where $W_f(f')$ is a weighting function that is zero for all f' except those that are close to f. This ensures that S(f) is the average of $|X(f)|^2$ at the neighbouring frequencies. More specific examples will be given later in Section 4.5.

Even when one has already chosen to make use of equation (4.7) as our spectral estimate, there are still at least three ways of **computing** this mathematically defined quantity. The obvious method is to actually compute X(f) and apply directly the definition, with integration replaced by summation for the sake of practicality. This is called the **frequency averaging method** or the **smoothed periodogram method**. The second method is to use tuned filters that will pick out from x(t) the part near frequency f and compute the power of that part. A third method, the **auto-correlation method** or **Blackman-Tukey method**, is more circumspect. To discuss this, we must first derive the **Wiener theorem**.

4.2.2 WIENER THEOREM

The Wiener theorem states that the power spectrum is the Fourier transform of the auto-correlation function. We have (again, $\omega = 2\pi f$)

$$S(f) = <|X(f)|^2> = <X(f)^*X(f)>$$

$$= < \int_{-\infty}^{\infty} \exp(j\omega t)x(t)dt \int_{-\infty}^{\infty} \exp(-j\omega s)x(s)ds>$$

$$= \iint_{-\infty}^{\infty} \exp\{j\omega(t-s)\}<x(t)x(s)>dtds \qquad (4.8)$$

We recognise $<x(t)x(s)> = R(|t-s|)$ (see Section 2.5.1) so that this becomes

$$S(f) = \iint_{-\infty}^{\infty} \exp\{j\omega(t-s)\}R(|t-s|)dtds$$

Putting $t-s = -\tau$ as a new integration variable, we see that the integrand is a function of τ only, or

$$S(f) = \int_{-\infty}^{\infty} \exp(-j\omega\tau)R(|\tau|)d\tau \qquad (4.9)$$

This equation has the corollary

$$R(|\tau|) = \int_{-\infty}^{\infty} \exp(j\omega\tau)S(f)df \qquad (4.10)$$

These are known as the Wiener−Khintchine relations, which permit us to calculate the power spectrum from the auto-correlation or vice versa.

We shall discuss in the next chapter how to compute R from values of x. For the moment, let us see how we can apply the sort of averaging operation one requires in equation (4.7) if we wish to use equation (4.9) to compute S(f). When actually computing S(f) by equation (4.7), one would not use a new $W_f(f')$ for every different value of f. This is neither feasible nor necessary. Instead, we would just use the same weighting function, but slide it along the frequency axis as f changes. This can be achieved simply by choosing

$$W_f(f') = W(f-f') \tag{4.11}$$

as in such a function a change to f just moves the whole function along the axis. Applying such a weighting operation to S(f') in equation (4.9) we have

$$\int W(f-f')S(f')df' = \int\int W(f-f')\exp(2\pi jf'\tau)R(\tau)d\tau df'$$

$$= \int\int W(f-f')\exp\{-2\pi j\tau(f-f')\}df'\exp(j\omega\tau)R(\tau)d\tau$$

Changing the variable of integration to f" = f-f' in the first integral, we have

$$LHS = \int\int W(f'')\exp(-2\pi jf''\tau)df'' \, \exp(j\omega\tau)R(\tau)d\tau$$

$$= \int w(\tau)R(\tau)\exp(j\omega\tau)d\tau \tag{4.12}$$

where we have defined $w(\tau)$ as the Fourier transform of W(f")

$$w(\tau) = \int W(f'')\exp(-2\pi jf''\tau)df'' \tag{4.13}$$

Equation (4.12) shows that taking an average on S using weighting function W(f-f') is equivalent to multiplying the Fourier transform of W, $w(\tau)$, into $R(\tau)$. Thus, the averaging operation can be performed in time space by a simple multiplication, rather than by taking moving averages in frequency space. The two are identical in effect.

There remains the question of what sort of weighting function W or w is the most suitable for a particular problem. This question will be discussed later in Section 4.5.

4.3   Cross-Spectra

A common requirement in signal analysis is to discover degrees of similarity between two signals. For example, the movements of two separate parts in a physical system subject to some form of random vibration may show the degree of vibrational transmissibility between the two components, and in this way, locate sources of resonance which may cause metal fatigue or other forms of serious damage. Such comparison of two signals can take place in the time domain or in the frequency domain. Time domain estimates are obtained by finding the mean of the cross-products of the two signals, point by point along the time axis. If the mean products are large, then we know that the signals contain a degree of similarity. By shifting the time axis of one of the signals relative to the other, the calculated mean product can be repeated at various time delays to discover whether any similarity exists between the two signals at different moments in time. This is the general idea of the method of cross-correlation to be discussed in the next chapter. A disadvantage of the method is that it tells us nothing about the relative phase characteristics of the two signals. This is illustrated in Fig. 4.1. Two composite signals A and B are shown, each derived from identical constituents in frequency and amplitude, and differing only in the relative phase of the two constituents. A cross-correlation of the type described above will clearly not identify any similarity. Yet, the two signals really contain the same two sinusoids, with the high frequency component at the same phase and the low frequency part at different phase shifts. To detect this sort of similarity, at the same time identifying phase information, requires measurement or analysis in the phase domain. This is the purpose of cross-spectrum analysis.

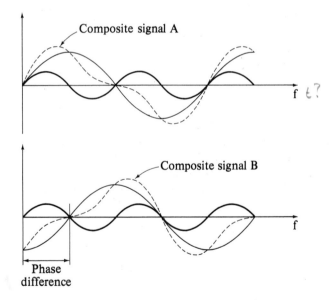

Fig. 4.1 Composition of two signals having identical constituents but differing in phase

An illustration of the method is shown in Fig. 4.2. Here a narrow 'slice' of both signals, A and B, is selected at a centre frequency f. The width of the slice is given as δf. The cross-product of the mean values of the signals over

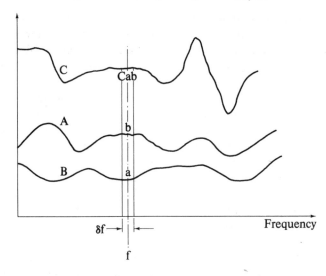

Fig. 4.2   Product of two signals in the frequency domain

the interval δf is obtained and shown as C_{ab} in the third curve. The process is repeated for different values of centre frequency f, to give the frequency product curve C. This is one form of the cross-spectrum of the two signals, obtained in the frequency domain. It will be effective in indicating pronounced similarity in the frequency domain and only to this extent will the technique be successful. A more generalised form of this analysis will achieve more than this; it will, in addition, identify phase characteristics.

To make the discussion more precise, we define the cross-spectrum of two random signals x and y as

$$S_{xy}(f) = <X^*(f)Y(f)> \tag{4.14}$$

To see its significance, we recall that the covariance of two random variables is

$$C(x,y) = <xy> - <x><y>$$

Now we have

$$<X^*(f)> = <\int x(t)\exp(j\omega t)dt>$$
$$= \int \exp(j\omega t)<x(t)>dt$$
$$= <x(t)>\int \exp(j\omega t)dt$$

This integral is 0 except for $f = 0$. Similarly for $<Y(f)>$. Thus, by definition, the covariance between $X^*(f)$ and $Y(f)$ is just $S_{xy}(f)$, except for the isolated case of $f = 0$. In this way, $S_{xy}(f)$ tells us something about how 'closely related' $X^*(f)$ and $Y(f)$ are.

However, there is a complication. X and Y may be complex, and S_{xy} would have a real and a complex part. How do we use them? It turns out that the more interesting quantities are the magnitude and the phase of S_{xy}. $|S_{xy}|$ increases with the size of both the real and the imaginary parts, and gives us an indication of the relation between $|X(f)|$ and $|Y(f)|$. The phase of S_{xy}, which can be computed by the formula $\arctan \frac{Re(S_{xy})}{Im(S_{xy})}$, indicates the average phase difference between $X^*(f)$ and $Y(f)$. An example of the two, called the **amplitude spectrum** and the **phase spectrum** respectively, is shown in Fig. 4.3.

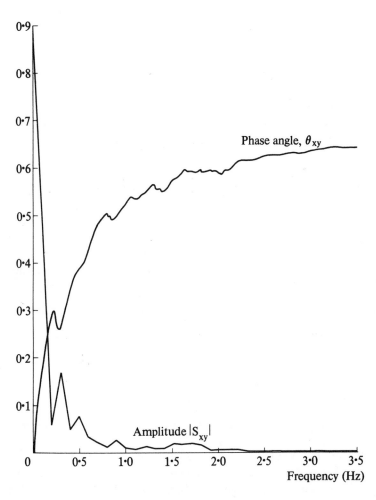

Fig. 4.3   Cross-spectral density

Another quantity derived from the cross-spectrum is the **coherence spectrum** (also called coherency spectrum) κ_{xy}

$$\kappa_{xy} = \frac{|\,S_{xy}(f)\,|^2}{S_x(f)S_y(f)} \qquad (4.15)$$

Here S_x is the power spectrum of x, and S_y that of y. The meaning of κ_{xy} is seen from the definition of correlation coefficient (Chapter 2) $\rho(x,y)$

$$\rho(x,y) = \frac{C(x,y)}{\sigma_x \sigma_y} \qquad (4.16)$$

The correlation coefficient between $X^*(f)$ and $Y(f)$ would be

$$\frac{<|\,X^*(f)Y(f)\,|>}{\{<|\,X(f)\,|^2><|\,Y(f)\,|^2>\}^{1/2}} = \frac{S_{xy}}{(S_x S_y)^{1/2}} \qquad (4.17)$$

However, because S_{xy} is a complex quantity, we prefer to take the absolute value to produce a real quantity. This gives us equation (4.15). In short, κ_{xy} plays the part of the correlation coefficient between $X(f)$ and $Y(f)$, showing how closely related they are. As in the case of ρ, the value of the coherence function cannot be greater than 1. If $\kappa_{xy}(f) = 0$ the two signals are said to be completely incoherent (that is, unrelated) at this frequency. If $\kappa_{xy}(f) = 1$, the two are fully coherent (or closely related) at f. It is important to realise that the estimates for S_x, S_y and S_{xy} must be obtained under identical conditions. (That is, the same filter bandwidth, record length, etc.) Otherwise gross errors in coherence estimation will be obtained.

Analogous to the Wiener theorem studied earlier, we have the result that the cross-spectrum is the Fourier transform of the cross-correlation. We have

$$<X^*(f)Y(f)> = \iint \exp(j\omega t)^* \exp(j\omega s) <x(t)x(s)> dt ds$$

$<x(t)x(s)>$, as we see in the next chapter, is the cross-correlation function $R_{xy}(t-s)$, and we have

$$S_{xy}(f) = \iint \exp\{-j\omega(t-s)\} R_{xy}(t-s) dt ds$$
$$= \int \exp(-j\omega\tau) R_{xy}(\tau) d\tau \qquad (4.18)$$

(Again, remember $\omega = 2\pi f$ and $\tau = t-s$.) The similarity of this with equation (4.9) should be obvious.

The estimation can again be done in various ways: (1) segmentation of x(t) and y(t), Fourier transform each segment and then average $X^*(f)Y(f)$; (2) Fourier transform the whole x and y, and take averages of $X^*(f)Y(f)$ for nearby frequencies; and (3) estimate $R_{xy}(\tau)$ first, multiply by $w(\tau)$ and finally Fourier transform. These are completely analogous to the methods for computing S_x.

4.4 Errors in Spectral Density Estimates

In spectrum estimation, errors can arise due to the inability to apply the complete features demanded by the mathematical method, such as our inability to integrate over an infinite time interval, inaccuracies of the method of data representation, such as the need to estimate with sampled data, and the imperfect nature of the acquisition itself. This latter factor relates to the equipment performance, e.g., amplifier non-linearity, frequency distortion, etc., and will not be considered here. Some of the factors upon which the reliability of the estimates will depend are: record length, filter bandwidth, filter shape, averaging time, scan rate, extraneous noise, sampling rate and quantisation levels. The most important of these are record length and filter bandwidth.

To discuss errors in more quantitative terms, we recognise three different entities related to the power spectrum from which to define the measurement error:

1. The true power spectrum $S(f)$, obtained from the theoretical model of the signal.
2. The measured value of the power spectrum, $\hat{S}(f)$, obtained over finite time for a particular piece of signal.
3. The mean-value of the measured spectrum, $<\hat{S}(f)>$, resulting from the averaging of a large number of measurements. For convenience we denote this as $\bar{S}(f)$.

As we discussed in Section 2.6, we measure the departure of $\hat{S}(f)$ from $S(f)$ by computing the mean-square deviation $<\{\hat{S}(f)-S(f)\}^2>$, which is naturally separated into two parts

$$<\{\hat{S}(f)-\bar{S}(f)\}^2> + \{\bar{S}(f)-S(f)\}^2$$

This is just a special form of equation (2.88). The first term is the variance of $\hat{S}(f)$, since it is the mean-square deviation from its own average value $\bar{S}(f)$. The second term is the bias of $\hat{S}(f)$, that is, the difference between the theoretical and the actual average values. We shall treat the two types of errors in the subsequent sections. But first we must discuss leakage.

4.4.1 LEAKAGE

A serious problem in Fourier analysis in general, and in spectral analysis in particular, is **leakage**, which is due to the inability of our analysis procedures to recognise harmonic components whose periods are not integer factors of our analysis interval; or, to put it another way, components whose frequencies are not integer multiples of our basic frequency $1/T$. A pure sinusoid of frequency f which is, say, somewhere between two integer frequencies i/T and $(i+1)/T$, would appear under Fourier transformation as a combination of integer frequency components. The signal power of this function is thus distributed, or 'leaked', to a wide range of frequencies. Thus, if the signal we analyse has a sharp peak at frequency f, under Fourier analysis its power is distributed over a set of neighbouring frequencies, and it appears as a series of peaks. The shorter is our analysis interval length T, the worse is the problem, while on the other hand, an infinitely long interval would eliminate leakage altogether.

To make the discussion more quantitative, let us consider the function $x(t) = \exp(2\pi jft)$. Suppose $f = i/T$, then when we compute the Fourier coefficients of $x(t)$ we would have

$$X_k = \frac{1}{T} \int_0^T \exp(2\pi jit/T)\exp(-2\pi jkt/T) = 1 \text{ for } k = i$$

$$= 0 \text{ for all other } k$$

Now suppose f is some other value. We then have

$$X_k = \frac{1}{T} \int_0^T \exp(2\pi jft)\exp(-2\pi jkt/T)$$

$$= \frac{1}{T} \int_0^T \exp\{2\pi j(ft-k)t/T\}dt$$

Carrying out the integration and substituting the integration limits we get

$$X_k = \frac{1}{T} \cdot \frac{\exp\{2\pi j(fT-k)\} - 1}{2\pi j(fT-k)/T}$$

$$= \frac{\exp\{\pi j(fT-k)\}}{\pi(fT-k)} \cdot \frac{\exp\{\pi j(fT-k)\} - \exp\{-\pi j(fT-k)\}}{2j}$$

We recognise the middle factor as the sine function (Euler's equation). Thus

$$X_k = \exp\{\pi j(fT-k)\}\frac{\sin\{\pi(fT-k)\}}{\pi(fT-k)}$$

and the signal power observed at frequency k/T would be

$$\mid X_k \mid^2 = \left| \frac{\sin\{\pi(fT-k)\}}{\pi(fT-k)} \right|^2$$

We see that the signal power is now distributed over all frequencies k/T, even though it is in reality confined to a single frequency f. The amount of power leaked to frequency k/T depends on the 'distance' between it and the 'true' frequency f, that is, on $T(f-\frac{k}{T})$. The closer k/T is to f, the more power it receives, while the more distant frequencies receive progressively less. The function that describes the leakage is frequently abbreviated to **sinc**, whose definition is

$$\text{sinc}(t) = \sin(\pi t)/(\pi t) \qquad (4.19)$$

With this notation, $\sin\{\pi(fT-k)\}/\{\pi(fT-k)\}$ can be re-written as sinc(fT−k), and we have the more concise notation for leakage

$$\mid X_k \mid^2 = \{\text{sinc}(fT-k)\}^2 \qquad (4.20)$$

Now let us take a general function x(t) and consider the relation between the computed Fourier coefficient X_k and the true Fourier transform X(f). We can express x(t) in terms of X(f) as

$$x(t) = \int_{-\infty}^{\infty} \exp(2\pi jft)X(f)df$$

By definition

$$X(f) = \frac{1}{T} \int_{0}^{T} \exp(-2\pi jkt/T) \int_{-\infty}^{\infty} \exp(2\pi jft)X(f)dfdt$$

Integrating over t first, the two exponential functions simply produce $\exp\{2\pi j(fT-k)\}$sinc(fT−k) as shown earlier. Thus, X_k is related to X(f) by the expression

$$X_k = \int_{-\infty}^{\infty} \exp\{2\pi j(fT-k)\}\text{sinc}(fT-k)X(f)df \qquad (4.21)$$

In other words, the 'finite' Fourier transform X_k is related to the 'infinite' Fourier transform by a weighted integration, the weighting function being just the sinc function.

The effect of this weighted integration is to mix together different frequencies in x(t). The sinc function takes maximum value at k = fT, so that X_k contains a large portion of the power at frequency f = k/T. However it also contains the power at other frequencies. This mixing, or leaking of power is due to, let us repeat, the fact that we are computing X_k by a finite interval integration, whereas $X(f)$ has been defined over an infinite interval.

A similar situation exists for spectral analysis. We know from equation (4.9) that $S(f)$ is the Fourier transform of $R(\tau)$ over $\{-\infty, \infty\}$. In actual computation we cannot produce an infinitely long $R(\tau)$, and must truncate it between, say, $-\hat{T}$ and $\hat{T}$ for some finite $\hat{T}$. (Note that $\hat{T}$ is usually much shorter than T, which is the length of the signal available for analysis. The two must be carefully distinguished.) The $\hat{S}(f)$ computed in this way is related to $S(f)$ by $\hat{S}(f) = \int \text{sinc}\{2T(f-f')\}S(f')df'$. In this way, leakage has the effect of producing spurious peaks in the computed spectrum, since the computed result is really the 'true' spectrum integrated with a sinc weighting function, which is highly oscillatory. These spurious peaks can be reduced by applying suitable averaging operations on the computed spectrum, which, as we shall see, also has the effect of increasing bias but reducing variance.

4.4.2 LEAKAGE AND SPECTRUM BIAS

By definition, $S(f) = <|X(f)|^2>$. Thus, if we simply use $|X(f)|^2$ as our computed spectrum $\hat{S}(f)$, then we would have an unbiased estimate, in other words, $\bar{S}(f) = S(f)$. However, the variance of such a spectrum would be very large. As we said in Section 4.2.1, it is essential to carry out some kind of averaging operation to reduce the variance, and one way of doing this is to smooth out $|X(f)|^2$ using a weighting function $W(f)$, or, we make

$$\hat{S}(f) = \int W(f-f')|X(f')|^2df' \qquad (4.22)$$

Immediately, we see that

$$\bar{S}(f) = <\hat{S}(f)> = \int W(f-f')<|X(f')|^2>df'$$
$$= \int W(f-f')S(f')df' \qquad (4.23)$$

Thus, $\bar{S}$ is no longer equal to S, but is instead a smoothed version of the latter. There is then a difference between the theoretical and the actual averages of $\hat{S}$; in other words, $\hat{S}$ is now a biased estimate. Clearly, the wider is the weighting function W, the greater is the difference because more divergent parts of S would be included in the integral. As a result, bias increases with the width of W. On the other hand, we shall see later that variance decreases when this happens. So the thing to do is to find the best compromise.

There is another source of bias in $\hat{S}$: When we do the calculation, we do not compute the real $|X(f)|^2$ since this requires integration over an infinite interval. What we have is an approximation to $|X(f)|^2$ computed over a finite interval. As we said in the last section, this produces leakage, which is mathematically equivalent to integrating $X(f)$ with the weighting function, $\text{sinc}(f-f')$. This effect is present implicitly. Now if we carry out another weighted integration on the spectrum using the function W, the two effects combine, and the total effect is an averaging operation with a weighting function which is a combination of W and sinc.

As we indicated earlier, by itself sinc is a poor weighting function. It is highly oscillatory, and converges slowly. It tends to introduce into the computed spectrum fictitious structures which are not really there. However, by carrying out a further averaging operation on the computed spectrum, we can produce a combined weighting function which is much better in shape, with faster convergence and smaller oscillations. Consequently, it is always necessary to carry out some smoothing operation on the computed spectrum, regardless of which computation algorithm we adopt. The process is usually called **windowing**, and a weighting function is called a window, something through which we 'look at' the spectrum. The spectrum we actually 'see' is never the true spectrum, as distortion caused by the window is always present. The point is to design the computation to replace bad windows like the sinc function by good windows. We shall discuss this further in the next section.

4.4.3 BIAS/VARIANCE TRADEOFF

An important concept in the discussion of bias/variance trade-off is the **bandwidth** of a window. It is, roughly speaking, the amount of averaging carried out by W on S. Unfortunately, it is somewhat difficult to define quantitatively bandwidth in a generally satisfactory sense. Some detailed discussion was given by Parzen [6]. Very approximately the bandwidth B is large if a window is flat and low, small if it is tall and thin. In this way, B gives an indication of how large a part of S is included in each value of $\hat{S}$.

While in general terms, a large B leads to a large bias, that is, a large value of $S - \bar{S}$, the exact amount depends on the structure of S, and cannot be estimated accurately beforehand, since the whole point of spectral analysis is trying to estimate S. It is somewhat easier to estimate the variance of $\hat{S}$ for a given B. When we have N sampled values of a function over duration T, we are able to compute the discrete Fourier transform $X(f)$ for $f = i/T$, $i = 0,1,\ldots,N/2$. (We note that the maximum frequency is $N/(2T) = 1/(2h)$, where h is the sampling interval. This is of course consistent with the sampling theorem, requiring $h = 1/(2f_N)$.) The frequency spacing between different values is $1/T$. Thus, a window with bandwidth B would be averaging over approximately $\frac{B}{1/T} = BT$ independent values of the Fourier transform. As we saw in Section 2.6, the variance of an estimate is inversely proportional to the number of independent values contained in it. Consequently, the variance of S is related to $(BT)^{-1}$. While we shall not go into details, one can in fact prove that the variance V and the bias β of S are as a first order of approximation

$$V(S(f)) \simeq S(f)^2/(BT)$$

and

$$\beta(S(f)) \simeq B^2 S''(f)/24 \qquad (4.24)$$

where $S''(f)$ is the second derivative of $S(f)$ [7]. Combining the two parts as in equation (2.88) we know that the mean-square deviation of $\hat{S}$ from S is

$$<\{\hat{S}(f) - S(f)\}^2> = \frac{S(f)^2}{BT} + \frac{B^4 S''(f)^2}{576} \qquad (4.25)$$

The product 2BT is often called the number of **degrees of freedom** of $\hat{S}$. It is supposed to measure, in an approximate way, how many independent parts are contained in each value of $\hat{S}(f)$. As we said earlier, there are BT discrete Fourier coefficients in $\hat{S}(f)$. However, each coefficient contains a real part and an imaginary part (except for $f = 0$ and $f = N/(2T)$) so that the number of independent parts is twice as large.

By taking averages using a window of width B, all the features in $S(f)$ less than B frequencies apart would be merged together, and we would not see them separately. Thus, the finer are the detailed structures of $S(f)$, the smaller we must choose B. This decreases the bias term in equation (4.25). However, the variance term increases. To compensate, we must increase T. This shows that the only way to see fine details in S is to have a long record. A short record enables us to see only the coarse structures. An example of this behaviour is shown in Fig. 4.4, where reductions in the bandwidth produce different results.

In a practical case we may know little about the signal before analysis and will need to develop some empirical way of determining the effective analysis bandwidth to use. One such technique is to carry out a preliminary analysis using a wide bandwidth (large number of degrees of freedom), and to repeat the analysis with progressively narrower bandwidths. If little change of shape occurs over a wide range of bandwidths, then a satisfactory resolution and accuracy will have been obtained. If the result of narrowing the bandwidth gives an increasingly fluctuating spectrum this could be due to insufficient record length, that is, number of discrete samples available. A fairly general experience is to find that the spectra will converge to a constant pattern as the filter bandwidth narrows until a given value is reached, beyond which a divergence of results will be obtained. The 'true' spectra can be taken at the filter bandwidth obtained just prior to this value. A suitable starting point for this procedure is found by examination of the auto-correlogram for the signal to be analysed. The delay τ is measured at the point where little significant change in the correlogram is seen and the filter bandwidth is taken as the reciprocal of this period.

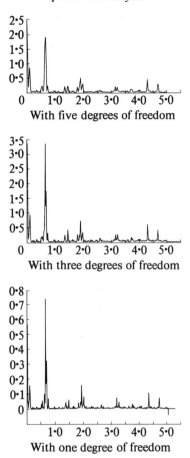

Fig. 4.4 Showing effect of different degrees of freedom

4.4.4 CONFIDENCE LIMITS*

In Section 2.6.3 we discussed the confidence limits of χ^2 variables. The applicability of the method of spectral analysis lies in the fact that, X(f) is the sum of $\exp(j\omega t)x(t)$ for all values of t, or, it is a linear combination of a large number of random variables. As we said in Section 2.4.2, through the central limit theorem, the sum of a large number of independent variables is approximately Gaussian. Now, $\hat{S}(f)$ is the average of BT values of $|X(f)|^2$. As explained a little earlier, it contains 2BT independent parts because each X(f) contains a real and an imaginary part. As we recall from Section 2.4.3, when we add up the square of n independent Gaussian variables the sum is a χ^2 variable with n degrees of freedom. In other words, we may analyse the behaviour of $\hat{S}(f)$ by considering it as a χ^2 variable with 2BT degrees of freedom, and discuss its confidence limits using well-known formulae and tables. As $\hat{S}(f)$ is an estimate of the spectral density, we can expect the true spectral value $S(f)$ to lie between two limits L_1 and L_2 with a probability of $100P$ per cent, i.e.

$$\text{Prob}\{L_1 \leqslant S(f) \leqslant L_2\} = P \qquad (4.26)$$

where

$$L_1 = S(f)(2BT)/\chi^2_{(2BT);\frac{1}{2}(1+P)} \qquad (4.27)$$

and

$$L_2 = S(f)(2BT)/\chi^2_{(2BT);\frac{1}{2}(1-P)} \qquad (4.28)$$

The two ratios $(2BT)/\chi^2_{(2BT);\frac{1}{2}(1\pm P)}$ are obtained from standard tables for χ^2 variables (for example, see reference [8]). The statistical tables available generally will only permit values up to about $2BT = 100$ to be obtained directly. Higher values can be computed from

$$\chi^2_{n;a} \simeq n + Z_a(2n)^{1/2} + \frac{2}{3}(Z_a^2 - 1) + \frac{1}{9}(2n)^{-1/2}(Z_a^3 - 7Z_a) \qquad (4.29)$$

where Z_a is the ordinate obtained from tables of the normal probability integral. Selected values of the upper and lower limit ratios, denoted as A and B respectively, for given values of confidence level P, are shown in Table 4.1.

A procedure for using confidence limits is as follows:

1. The number of degrees of freedom for an estimate is determined from record length T, bandwidth B for a particular spectral window (to be discussed later) found from Table 4.2, and the relation n = 2BT.
2. The confidence level is decided, e.g. P = 90 per cent, 80 per cent, or whatever one chooses.
3. Confidence parameters A and B are determined from Table 4.1.
4. Each value of $\hat{S}(f)$ is multiplied by A and B in turn to determine the accuracy limits at each frequency f.

To take an example, it is required to estimate the power in a spectrum for a record length T = 1s, and which is band-limited to contain no frequencies higher than 2 kHz. It is assumed that the width of the narrowest peak in the spectrum will be no greater than 40 Hz. The data are to be smoothed before spectral estimation by a Hanning window. The sampling rate must be ≥4 kHz and the sampling interval ≤0·25 ms. Hence the number of degrees of freedom will be obtained from n = 2BT and modified by the bandwidth of the equivalent smoothing filter using Table 4.2 to obtain n = 2 x 40 x 1·3 = 100. For a confidence level of 90 per cent the modifying values are A = 1·28

Table 4.1 **Table of confidence limits for auto-spectrum evaluation**

N	P = 99%		P = 90%		P = 80%		P = 50%	
	A	B	A	B	A	B	A	B
1	25641·0256	0·1269	256·4103	0·2604	63·2911	0·3690	9·4340	0·7364
2	200·0000	0·1887	19·4175	0·3339	9·4787	0·4338	3·4542	0·7994
3	41·8410	0·2336	8·5227	0·3841	5·1370	0·4800	2·4712	0·7184
4	19·3237	0·2692	5·6259	0·4215	3·7736	0·5141	2·0812	0·7362
5	12·1359	0·2985	4·3478	0·4517	3·1056	0·5411	1·8720	0·7490
6	8·8757	0·3235	3·6585	0·4766	2·7273	0·5639	1·7396	0·7649
7	7·0779	0·3452	3·2258	0·4975	2·4735	0·5842	1·6486	0·7698
8	5·9701	0·3643	2·9304	0·5158	2·2923	0·5988	1·5810	0·7784
9	5·2023	0·3815	2·7027	0·5319	2·1583	0·6131	1·5291	0·7860
10	4·6296	0·3970	2·5381	0·5461	2·0534	0·6254	1·4874	0·7930
20	2·6918	0·5000	1·8433	0·6367	1·6077	0·7040	1·2968	0·8366
30	2·1755	0·5590	1·6225	0·6854	1·4563	0·7452	1·2277	0·8598
40	1·9314	0·5991	1·5089	0·7174	1·3769	0·7721	1·1902	0·8749
50	1·7848	0·6290	1·4382	0·7406	1·3274	0·7912	1·1665	0·8861
60	1·6875	0·6525	1·3892	0·7587	1·2921	0·8061	1·1493	0·8944
70	1·6168	0·6717	1·3529	0·7732	1·2657	0·8181	1·1363	0·9011
80	1·5628	0·6878	1·3247	0·7852	1·2451	0·8280	1·1261	0·9066
90	1·5199	0·7015	1·3020	0·7954	1·2285	0·8364	1·1178	0·9112
100	1·4849	0·7135	1·2832	0·8042	1·2147	0·8436	1·1109	0·9152
200	1·3136	0·7836	1·1885	0·8547	1·1442	0·8847	1·0752	0·9377
300	1·2464	0·8178	1·1500	0·8787	1·1152	0·9040	1·0603	0·9483
400	1·2087	0·8393	1·1279	0·8936	1·0985	0·9159	1·0517	0·9548
500	1·1839	0·8544	1·1132	0·9039	1·0873	0·9242	1·0459	0·9593
600	1·1661	0·8659	1·1026	0·9117	1·0792	0·9304	1·0417	0·9626
700	1·1524	0·8749	1·0944	0·9178	1·0729	0·9353	1·0384	0·9653
800	1·1416	0·8823	1·0879	0·9228	1·0680	0·9392	1·0358	0·9674
900	1·1328	0·8885	1·0826	0·9270	1·0639	0·9425	1·0337	0·9692
1000	1·1254	0·8937	1·0781	0·9305	1·0604	0·9453	1·0319	0·9707

A = coefficient of upper limit, B = coefficient of lower limit, P = percentage of auto-spectrum value between these limits, N = degrees of freedom for spectral evaluation.

and B = 0·78, so that we need to multiply $\hat{S}(f)$ by these values to determine the error band. For a given percentage probability, the width of the band will be dependent on the instantaneous value of S(f) at any frequency. If the spectral estimates are plotted on a logarithmic scale, the confidence interval for the spectral estimate may be indicated as a constant vertical ordinate applicable to any part of the spectrum. That is, if we take the logarithm of each side of equations (4.27) and (4.28), the limits L_1 and L_2 will become

$$\text{Log } \hat{S}(f) + \log(2BT) - \log \chi^2_{(2BT);\frac{1}{2}(1 \pm P)} \qquad (4.30)$$

TABLE 4.2 Characteristics of smoothing windows

Type	Lag window ($W(\tau)$)	Spectral window ($W(f)$)	Equivalent bandwidth (B_s)
Rectangular	$= 1$ for $|\tau| \leqslant \hat{T}$ $= 0$ for $|\tau| > \hat{T}$	$= 2\hat{T} \left[\dfrac{\sin 2\pi f\hat{T}}{2\pi f\hat{T}} \right] = W_0$	$\dfrac{0\cdot 5}{\hat{T}}$
Bartlett	$= 1 - \dfrac{|\tau|}{\hat{T}}$ for $|\tau| \leqslant \hat{T}$ $= 0 \qquad$ for $|\tau| > \hat{T}$	$= \hat{T} \left[\dfrac{\sin \pi f\hat{T}}{\pi f\hat{T}} \right]^2$	$\dfrac{1\cdot 5}{\hat{T}}$
Parzen	$= 1 - 6\left(\dfrac{|\tau|}{\hat{T}}\right)^2 + 6\left(\dfrac{|\tau|}{\hat{T}}\right)^3$ for $|\tau| \leqslant \dfrac{\hat{T}}{2}$ $= 2\left(1 - \dfrac{|\tau|}{\hat{T}}\right)^3$ for $\dfrac{\hat{T}}{2} < |\tau| \leqslant \hat{T}$ $= 0 \qquad\qquad$ for $|\tau| > \hat{T}$	$= 0\cdot 75\hat{T} \left[\dfrac{\sin \pi f\,{}^1\!/_2\hat{T}}{\pi f\,{}^1\!/_2\hat{T}} \right]^4$	$\dfrac{1\cdot 9}{\hat{T}}$
Hanning	$= 0\cdot 5 + 0\cdot 5 \cos\left(\dfrac{\pi\tau}{\hat{T}}\right)$ for $|\tau| \leqslant \hat{T}$ $= 0 \qquad\qquad\qquad$ for $|\tau| > \hat{T}$	$= 0\cdot 5 W_0(f)$ $+ 0\cdot 25 W_0\left[f + \dfrac{1}{2\hat{T}}\right] + 0\cdot 25 W_0\left[f - \dfrac{1}{2\hat{T}}\right]$	$\dfrac{1\cdot 3}{\hat{T}}$

TABLE 4.2 Characteristics of smoothing windows (continued)

Type	Lag window ($W(\tau)$)	Spectral window ($W(f)$)	Equivalent bandwidth (B_s)
Hamming	$= 0\cdot54 + 0\cdot46 \cos\left(\dfrac{\pi\tau}{\hat{T}}\right)$ for $|\tau| \leqslant \hat{T}$ $= 0$ for $|\tau| > \hat{T}$	$= 0\cdot54 W_0(f)$ $+0\cdot23 W_0\left[f + \dfrac{1}{2\hat{T}}\right] + 0\cdot23 W_0\left[f - \dfrac{1}{2\hat{T}}\right]$	$\dfrac{1\cdot3}{\hat{T}}$
Blackman	$= 0\cdot42 + 0\cdot5 \cos\left(\dfrac{\pi\tau}{\hat{T}}\right)$ $+0\cdot08 \cos\left(\dfrac{2\pi\tau}{\hat{T}}\right)$ for $|\tau| \leqslant \hat{T}$ $= 0$ for $|\tau| > \hat{T}$	$= 0\cdot42 W_0(f) + 0\cdot25 W_0\left[f + \dfrac{1}{2\hat{T}}\right]$ $+0\cdot25 W_0\left[f - \dfrac{1}{2\hat{T}}\right] + 0\cdot04 W_0\left[f + \dfrac{1}{\hat{T}}\right]$ $+0\cdot04 W_0\left[f - \dfrac{1}{\hat{T}}\right]$	$\dfrac{1\cdot4}{\hat{T}}$

This permits the confidence interval simply to be added to the estimate obtained. An example is given in Fig. 4.5, which indicates the analysis bandwidth by means of a short horizontal line and the accuracy by means of a series of vertical ordinates for different confidence levels. The presentation of the power spectrum in logarithmic form is common in communication engineering, where signal levels are measured in decibels, which indicate the proportional change of power. A different way of handling confidence limits is discussed by Wonnacott [9].

4.5   Smoothing Functions

4.5.1   TIME AND FREQUENCY WINDOWS

Let us summarise what we have learnt so far. We can estimate $S(f)$ by computing $|X(f)|^2$ first and performing some frequency averaging over it as given by equation (4.22). Alternatively, we can do it by time averaging, either by segmenting $x(t)$ into time slices and averaging $|X(f)|^2$ over these, or by first estimating $R(\tau)$ from $x(t)$ and apply the Wiener theorem. However, time averaging alone is not enough; we must always apply some frequency averaging as well because the computed $|X(f)|^2$ contains leakage, which is suppressed by frequency averaging. Finally, it is possible to perform frequency averaging in time space as shown by equation (4.12).

We have derived all our formulae in terms of integrals over a continuous frequency f. In computation, we must use discrete transforms, and integrals are approximated by sums. The general, and somewhat surprising, rule followed is that f must be integer multiples of $(2\hat{T})^{-1}$, where $\hat{T}$ is the interval length over which we perform Fourier transformation. In other words, if we slice the input signal $x(t)$ into segments of length $\hat{T}$ in the segment averaging method, then we must compute Fourier coefficients at $f = i/2\hat{T}$. If, on the other hand, we prefer to use the auto-correlation method, and want to truncate $R(\tau)$ between $\tau = 0$ and $\tau = \hat{T}$, then again we must compute Fourier coefficients at $f = i/2\hat{T}$.

The rule may appear to be inconsistent with the discussion of Chapter 3, where we said that, when performing a DFT over interval $\{0,T\}$ we should compute the frequencies i/T, $i = 0,1,\ldots,N-1$. To explain this discrepancy, let us note that equation (4.9) is a **symmetric** integral between $\pm\infty$. Now quite obviously it is not feasible to compute $R(\tau)$ for infinitely large values of τ. So instead, in equation (4.11) we would choose a window $w(\tau)$ that is zero outside the interval $\{-\hat{T},\hat{T}\}$, so that values of $R(\tau)$ for larger τ are eliminated from the expression and need not be computed. This makes $S(f)$, the Fourier transform of $R(\tau)$ over $\{-\hat{T},\hat{T}\}$, which is an interval of length $2\hat{T}$. Accordingly, the frequencies we need to be concerned with are f $= i/(2\hat{T})$, consistent with the discussion of Chapter 3. However, because

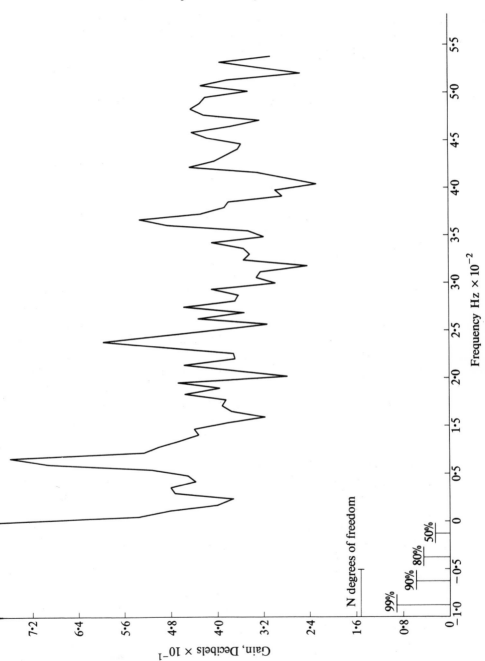

Fig. 4.5   Calibrated plot of power spectral density on a logarithmic scale

$R(\tau)$ is symmetrical, we only need to find its values over $\{0,T\}$. Thus, we *appear* to be working with an interval of length $\hat{T}$.

The above discussion is based on the Wiener theorem. However, the rule concerning frequency values is equally valid for other spectrum computation methods, because mathematically they are all equivalent. It is the failure to follow this rule that leads to some difficulty concerning the correct windowing procedure for the segment averaging method. (See the following subsection.) If given x(t), we simply divide it into segments of length $\hat{T}$ and perform an FFT over each segment, we could only compute S(f) at $f = i/\hat{T}$, and it would not be possible to carry out the required windowing operation because half of the needed terms are missing. The correct procedure is to extend each segment to length $2\hat{T}$, by attaching to each block of x a block of zeros. Performing an FFT on the enlarged segment then produces Fourier coefficients at frequencies $i/2\hat{T}$.

We have said that, no matter which computation algorithm we choose, we are effectively truncating $R(\tau)$ between $\pm\hat{T}$. The computed spectrum and the true spectrum, then, are related via the leakage weighting function for an interval of length $2\hat{T}$. This function is $2\hat{T}\text{sinc}(2f\hat{T})$. In addition, we impose a smoothing window of some sort to suppress leakage and reduce variance, which is equivalent to multiplying some windowing function $w(\tau)$ into $R(\tau)$. The combined effect is that $R(\tau)$ becomes $R(\tau)w(\tau)$ for $-\hat{T}\leqslant\tau\leqslant\hat{T}$ and 0 outside the interval. The corresponding effect in frequency space is that S(f) has been smoothed out by a windowing function W(f), which is the Fourier transform of $w(\tau)$ over $\{-\hat{T},\hat{T}\}$

$$W(f) = \int_{-\hat{T}}^{\hat{T}} \exp(2\pi jf\tau)w(\tau)d\tau \tag{4.31}$$

Our job in spectral analysis is to select a $w(\tau)$ such that the W(f) corresponding to it is free of oscillations (so that it would not introduce fictitious peaks into the computed spectrum) and has a sharp main peak that quickly decays to zero as f moves away from the origin (so that it would not unduly merge the features of the spectrum and make them unidentifiable). There are several possible choices for $w(\tau)$; none of these can be termed best in a theoretical sense but the last one is decidedly the most popular:

1. Bartlett window:

$$w(\tau) = 1 - |\tau|/\hat{T}, \quad |\tau|\leqslant\hat{T}$$
$$= 0 \qquad\qquad |\tau|>\hat{T} \tag{4.32}$$

If we substitute this into equation (4.31) we would get

$$W(f) = \hat{T}(\frac{\sin\pi f\hat{T}}{\pi fT})^2 \tag{4.33}$$

2. Parzen window:

$$w(\tau) = 1 - 6(\tau/\hat{T})^2 + 6(|\tau|/\hat{T})^3, |\tau| \leqslant \tfrac{1}{2}\hat{T}$$

$$= 2(1-|\tau|/\hat{T})^3, \qquad \tfrac{1}{2}\hat{T} < |\tau| \leqslant \hat{T} \qquad (4.34)$$

$$= 0 \qquad\qquad\qquad |\tau| > \hat{T}$$

Its formula in frequency space is

$$W(f) = \frac{3}{4} \hat{T} \left(\frac{\sin^{1/2}2\pi f\hat{T}}{{}^{1/2}\pi fT} \right)^4 \qquad (4.35)$$

3. Hanning window:

$$w(\tau) = \tfrac{1}{2} + \tfrac{1}{2}\cos(\pi\tau/\hat{T}), \quad |\tau| \leqslant \hat{T}$$

$$= 0 \qquad\qquad\qquad |\tau| > \hat{T} \qquad (4.36)$$

$$W(f) = \hat{T}\sin(2\pi f\hat{T})/\{2\pi f\hat{T}[1-(2f\hat{T})^2]\} \qquad (4.37)$$

Finally, if we do not impose any smoothing window at all, this is equivalent to simply truncating $R(\tau)$, which is described as **rectangular** windowing. Its $W(f)$ is just $2\hat{T}\mathrm{sinc}(2\hat{T}f)$. Table 4.2 summarises the properties of the main windows. Their bandwidths are shown, though we do not discuss how they are defined or derived. Note that B is inversely related to $\hat{T}$. Detailed analysis may be found in [10]. Figure 4.6 shows the four windows in time and frequency spaces.

While theoretical considerations may be advanced for the superiority of each window, in practice there is little difference in their performance as long as we choose $\hat{T}$ carefully to make their bandwidths equal. The reason for the popularity of the Hanning window is that it is equally easy to use in either time or frequency spaces. In time space, we multiply w into R as in equation (4.12); in frequency space, we use the formula (4.23), a weighted integration. But in actual computation we perform summation, not integration. Since $w(\tau)R(\tau)$ is truncated between $\pm\hat{T}$, we should use the frequencies $f = i/(2\hat{T})$. Thus, the computed spectrum should be the following discrete version of equation (4.22)

$$\hat{S}(f) = \sum_k W(f - \frac{k}{2T}) \; |X(\frac{k}{2T})|^2 \qquad (4.38)$$

with $f = i/2\hat{T}$. Now if we compute $W(\frac{i-k}{2T})$ for the Hanning window, we find that

$$W(\frac{i-k}{2T}) = \tfrac{1}{2}, k = i$$

$$= \tfrac{1}{4}, k = i \pm 1$$

$$= 0 \text{ for all other values of } k \qquad (4.39)$$

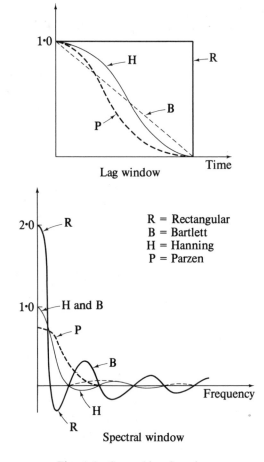

Fig. 4.6   Smoothing functions

Thus, equation (4.38) is now simplified into a sum of just three terms

$$\hat{S}(\frac{i}{2T}) = \frac{1}{4}|X(\frac{i-1}{2T})|^2 + \frac{1}{2}|X(\frac{i}{2T})|^2 + \frac{1}{4}|X(\frac{i+1}{2T})|^2 \tag{4.40}$$

This is a special property of the Hanning window. The other windows do not have a simple W like equation (4.39), and computation of $\hat{S}$ does not have a simple frequency space summation like equation (4.40). In this way, the Hanning window is easy to apply in either space, while Bartlett and Parzen windows can be conveniently used only in time space by applying equation (4.11).

The importance of this point becomes even clearer when we consider windowing in the segmentation method. If we slice x(t) into segments of length $\hat{T}$, the procedure is theoretically equivalent to the auto-correlation method in which $R(\tau)$ is truncated between $\pm\hat{T}$. In both cases, we obtain $\hat{S}(f)$ at $f = i/2\hat{T}$. However, in the latter we can choose to apply windowing on $R(\tau)$ by multiplying it into $w(\tau)$; in the former method we never actually compute R, and this windowing method is simply not applicable. The only thing we can do is to perform windowing in frequency space by the use of equation (4.40). Therefore, in the segmentation method the Hanning window is in fact the only window we can use conveniently. The other two windows, because of their complex frequency space expression, are easy to use only in time space, and cannot be applied in the segmentation method.

The situation is different yet again in the frequency averaging method. Here we Fourier transform the whole x(t), which is of duration T rather than $\hat{T}$. Now that the Fourier transformation is performed over interval T, we should compute $X(f)$ at $f = i/2T$. That is, with a much longer interval we would space f much closer. Averaging is performed by adding up many neighbouring frequencies, as specified in equation (4.23). In the other methods, it is necessary to add up only a few neighbouring frequencies, because in those cases some averaging has already been achieved in other ways. In the auto-correlation method $R(\tau)$ is an average quantity, while in the segment averaging method $| X(f)|^2$ is averaged over many segments. In either case, only a small amount of frequency averaging is needed to suppress leakage and produce a good window shape. In the frequency averaging method adding up neighbouring frequencies is the only averaging we perform, and the amount must be correspondingly larger. However, here we have the slight advantage that we need to apply only one sort of averaging to achieve two aims: leakage suppression and reduction of variance.

A similar situation exists for the filtering method. In this, a filter that passes only certain frequencies and rejects others is used and its output is squared and integrated to produce the signal power in the frequency band. It should be clear that this automatically sums the power of many neighbouring frequencies and achieves the same aim as frequency averaging.

4.5.2 COMMENTS ON MODERN METHODS FOR SPECTRAL ESTIMATION

All the windowing methods we mentioned so far apply **quadratic** windows because they operate on second-order quantities, whether $R(\tau)$, which is $<x(t)x(t+\tau)>$, or $| X(f)|^2$. In recent years it has become increasingly popular to perform **linear** windowing as a means of leakage suppression. In this, a window w(t) is multiplied into x(t), either all of it, or its segments, before either is Fourier transformed. We advise against this method. Some of the reasons are given in a paper by Sloan [11] and will not be repeated here. Listed below are several additional reasons.

Our first point is simply that linear windowing is unnecessary. The function it is supposed to perform, leakage suppression, is already quite adequately achieved through quadratic windowing. Regardless of what method we use, suitable quadratic windowing will always smooth out the oscillations and give a good overall weighting function W(f−f′).

Linear windowing is also unsatisfactory from several theoretical points of view. First, it does not conserve power. That is, multiplying any common linear window into the signal function (1,0,0,...,0,1) will produce an identically zero function, with zero power. Quadratic windowing will always conserve power if we choose w(0) = 1, because we know that by equation (4.6)

$$\int S(f)df = R(0) \qquad (4.41)$$

If we choose w(0) = 1 and multiply w into R, we do not change R(0), and we still have $\int \hat{S}(f)df = R(0)$, so that the total power in S is reproduced in $\hat{S}$. This does not hold in the linear window case because w(t) multiplies into x(t) and not R.

As even the proponents of linear windowing realise, the procedure effectively 'throws data away' by giving some of the data points small weighting factors. It is well known in statistics that the weighting factor given to an individual data point should be inversely related to its variance. That is, the more uncertainty there is in the individual value, the less weight it should have. It is then clear that, when we process a time-invariant random process, whose elements have a constant variance, each data point should be given equal weight. Linear windowing clearly violates this concept. In contrast, in quadratic windowing every procedure corresponds to the multiplication of a window into the auto-correlation function, whose elements have small variance near the origin and large variance away from it. Since all the windows have maximum value of one at the origin and decrease as we move away from it, the rule of statistical estimation is being satisfied.

We then see why many recent authors have recommended the use of 'overlapping' in the segment averaging method for spectrum estimation. Because the linear window gives the start and the end of each segment small weights, it becomes necessary to use the elements twice. That is, segment two does not start where segment one ends. Instead, its beginning part overlaps the end of the previous segment, so that those elements which have been reduced by the window are used twice, giving them a larger weighting factor.

It seems to us, however, that this is a rather untidy way of overcoming the weighting factor problem. The same aim can be achieved simply by doing away with linear windows altogether, since quadratic windows are just as effective for the purpose of leakage suppression. Furthermore, we have seen no theoretical or experimental demonstration that the computed results are any better. Few of the arguments in favour of the method went further than that 'this looks like a good idea'.

In terms of cost, in the segment averaging method with linear windowing, there is the need to multiply the window into each segment; this is eliminated in the method we recommend (see next section), though there is the extra cost incurred by the requirement of extending the interval to twice its given length and performing an FFT over the doubled interval. However, once overlapping is included the cost of linear windowing is further increased because there are now more segments. Taking both factors into account, there seems to be no reason to believe that linear windowing would save computing cost. But what about performance then? In Section 4.6.5 and in reference [12] we compare results computed by quadratic windowing with those obtained using linear windowing with no overlapping. Unless overlapping produces a substantial improvement for linearly windowed spectral estimates, a proposition for which there is no concrete proof, we see no reason to recommend linear windows.

Second, with linear windowing, equation (4.23) is no longer valid, and there is no simple relation between S and Ŝ on average. In other words, we are not even sure what we are actually estimating once we apply linear windowing. Finally, in our computed results, shown in Section 4.7, it will be seen that linearly windowed estimates are not as good as quadratically windowed ones. In our view linear windowing is thus unsatisfactory in theory and unnecessary in practice.

4.6 Spectral Analysis Summary

Most of the methods for spectral analysis have been discussed in the previous sections; in this section we provide short summaries only. However, the filtering method, being rather different, has not received much attention earlier, and so will be explained in detail here. In all cases we assume that there are N sampled values of x(t) input to the process.

4.6.1 THE SEGMENT AVERAGING METHOD

1. Divide the input into segments of M values each. The ratio N/M should be around 16 for usual cases. ($2N/M$ is approximately the number of degrees of freedom in the spectral estimate.)
2. Attach M zeros to each segment and perform a 2M-point FFT. This produces X(f) at $f = i/2\hat{T}$ as required. (If we take only an M-point FFT on each segment we would have X(f) for $1/\hat{T}$ only, and that would not be enough.)
3. Average $|X(f)|^2$ over all the segments.
4. Apply equation (4.38) to produce the computed spectrum Ŝ(f).

5. If $S_{xy}(f)$ is required for two inputs $x(t)$ and $y(t)$, follow **(1)** and **(2)** for both x and y; then average $X(f)*Y(f)$ over the segments, and apply (4.38) as before.

4.6.2 THE PERIODOGRAM APPROACH

1. Add N zeros to $x(t)$ and perform a 2N-point FFT.
2. Take averages of neighbouring values in $|X(f)|^2$. The number of points to include in each average should be around 32, which is approximately the number of degrees of freedom.
3. If S_{xy} is required, simply average over $X(f)*Y(f)$.

4.6.3 THE AUTO-CORRELATION APPROACH

1. Compute the mean of $x(t)$ and subtract it from each value of x.
2. Estimate R by computing the average

$$R_i = \frac{1}{N} \sum_{k=0}^{N-i-1} x_k x_{k+i}, \quad i = 0,1,\ldots,M \tag{4.42}$$

Again $2N/M$ is approximately the degree of freedom, and should be around 32. Divide R_0 by 2.

3. Multiply R_i by a window extending from $w_0 = 1$ to $w_{M+1} = 0$.
4. Compute
$$S(k) = \frac{1}{M} \sum_{i=0}^{M} \cos(\pi ki/M) w_i R_i \tag{4.43}$$
5. If the Hanning window is used, the alternative to **(3)** and **(4)** is

First compute $\quad S'(k) = \frac{1}{M} \sum_{i=0}^{M} \cos(\pi ki/M) R_i \tag{4.44}$

Then perform windowing
$$S(k) = \tfrac{1}{4}S(k-1) + \tfrac{1}{2}S(k) + \tfrac{1}{4}S(k+1) \tag{4.45}$$
$S(k)$ being $S(f)$ for $f = k/2T$
6. If S_{xy} is required, compute

$$R_{xy}(i) = \frac{1}{N} \sum_{k=0}^{N-i-1} x_k y_{k+i} \quad i \geq 0$$

$$= \frac{1}{N} \sum_{k=-i}^{N} x_k y_{k+i} \quad i < 0 \tag{4.46}$$

Then compute

$$S_{xy}(k) = \frac{1}{2M} \sum_{i=-M}^{M} \exp(\pi ki/M)w_i R_{xy}(i) \qquad (4.47)$$

or alternatively, compute

$$S'_{xy}(k) = \frac{1}{2M} \sum_{i=-M}^{M} \exp(\pi ki/M)R_{xy}(i) \qquad (4.48)$$

followed by

$$S_{xy}(k) = \frac{1}{4}S'_{xy}(k-1) + \frac{1}{2}S'_{xy}(k) + \frac{1}{4}S'_{xy}(k+1) \qquad (4.49)$$

4.6.4 USE OF A DIGITAL FILTER

A fundamental method of spectral analysis consists of passing the signal through a band-pass filter to determine the amplitude coefficients and squaring and averaging the output to provide a frequency point for a power spectral density analysis. Either the filter centre frequency can be changed after each measurement to obtain a complete spectral estimation or a contiguous approach carried out. The method can be expressed as

$$S_f = \frac{1}{N} \sum_{i=0}^{N} \left\{ \sum_{k=0}^{M} h_k x_{i+k} \right\}^2 \qquad (4.50)$$

where h_k is the impulse response function of a band-pass filter comprising M points with centre frequency f. This expression represents the time average of the squared convolution integral for the filtered data. The bracketed expression will be recognised as an FIR filter. In practical cases an IIR filter would be chosen as this enables a reduction in analysis time to be achieved. Of these the Lerner filter is particularly suitable and has an advantage in contiguous operation since some of its poles can be shared by adjacent filters [13]. It has also an equiripple amplitude characteristic and linear phase characteristics. This filter may be considered [14] in terms of the summation of a number of simpler subfilters consisting of a complex pole and a single zero (Fig. 4.7). Each of these subfilters is characterised by a complex pole-pair at the same radius but a different angle in the z-plane, so that they can be considered as individual narrow-band filters, each covering a section of the pass-band. The composite band-pass filter characteristics is obtained by adding and subtracting alternative subfilter outputs. Cut-off rates of some 60 db/octave are obtained with as few as eight subfilters of this type. Further discussion of the design of IIR filters is given in Chapter 6.

The use of constant bandwidth filters suffers from the disadvantages of unequal ratios of centre frequency/resolution bandwidth over the total analysis band, and a filter setting time which limits attainable scanning rate.

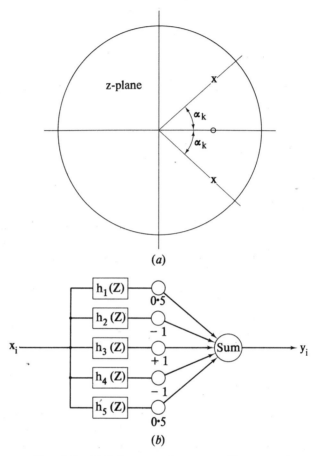

Fig. 4.7   (a) Pole-zeros of a Lerner filter
           (b) Representation of the Lerner filter

These effects render constant bandwidth filter methods inferior to FFT methods for values of N greater than about 50. However, the position is altered if a proportional bandwidth filter is used. A simple method based on the decimation technique is available for changing the filter bandwidth.

Referring to Section 6.2.2 if the sampling interval remains constant, the effective filter width can be halved by simply discarding every other point. Enochson and Otnes [15] have used this technique effectively to produce a third octave analysis algorithm which is faster than the FFT for values of N less than 40,000. Four IIR filters need to be implemented. Three of these are arranged each to cover one octave of the signal to be analysed and a fourth to carry out low-pass filtering before decimation (Fig. 4.8). Following each decimation by two the filter will react on the data at frequencies

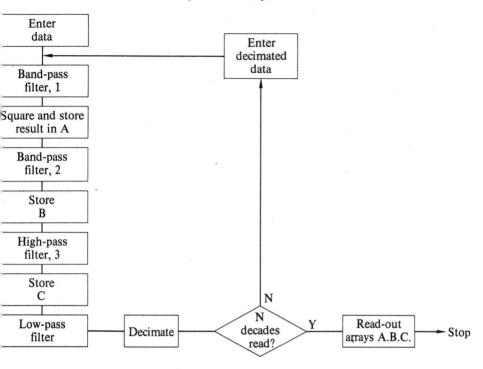

Fig. 4.8   Spectral analysis for recursive filtering and decimation

one octave lower to produce the next set of $\frac{1}{3}$ octave values and so on. The time required for analysis is the summation of a geometric series since each successive operation on the data will be carried out on half the preceding number of points.

4.6.5   SPECTRUM ANALYSIS EXAMPLES

In this section we illustrate the earlier discussion with computed examples. We shall compare the performance of different computing procedures by applying them to the same set of input functions and examining the resulting spectra. The input data are artificially generated by producing a set of Gaussian random numbers using a random number generator [16] and then filtering them in several ways to produce x(t) with known S(f), which can be compared with the computed $\hat{S}(f)$ for each method.

We will use N = 512. The set of random numbers is called n_k. Two different x series are produced by

1. $x_k = \frac{1}{4}n_{k-1} + \frac{1}{2}n_k + \frac{1}{4}n_{k+1}, \ k = 1,2,\ldots,510$

$x_0 = \frac{3}{4}n_0 + \frac{1}{4}n_1, \ x_{511} = \frac{3}{4}n_{511} + \frac{1}{4}n_{510}$

2. $x_0 = n_0$, $x_1 = n_1$, $x_k = x_{k-1} - \frac{1}{2}x_{k-2} + n_k$, $k = 2,3,\dots,511$

They have the following theoretical spectra

1. $S_i = \frac{1}{2}\{1+\cos(2\pi i/N)\}$

2. $S_i = \{2\frac{1}{4}-3\cos(2\pi i/N)+\cos(4\pi i/N)\}^{-1}$

These may be derived easily using the methods shown in Section 4.7.1.

Figures 4.9 and 4.10 show twelve spectra computed from these pieces of data using six different methods.

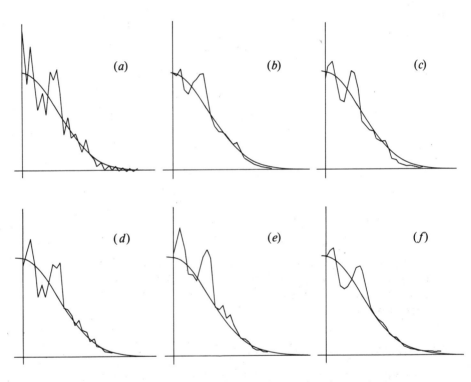

Fig. 4.9   Comparison of six spectrum estimation methods

(a)   the first 32 values of R were computed and transformed without windowing in accordance with equation (4.44).

(b)   the same spectra were smoothed applying equation (4.45).

(c)   the smoothed periodogram method was used with 32 point averages taken from a 512 point transform — squared spectrum, $|X(f)|^2$.

(d)   the segment averaging method was used with x(t) divided into 16 segments and $|X(f)|^2$ is applied.

(e)   a linear window was multiplied into each segment of x(t) *before* X(f) is computed.

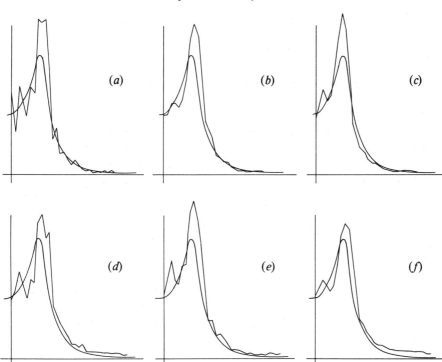

Fig. 4.10   Second comparison of spectrum estimation methods

(f)   equation (4.38) is applied to $|X(f)|^2$ after segment averaging.

There seems little doubt that **(b)**, **(c)** and **(f)** are superior to the others. In all three we carried out quadratic windowing, which produce better results than linear windowing **(e)**, or no windowing **(a)** and **(d)**.

The statistically minded reader will point out that the six spectra in each figure have unequal degrees of freedom, and question whether it is correct to compare them. As we said in Section 4.4.2, increasing the degree of freedom of a spectrum reduced variance but increases bias. However, we are not comparing the two individually. Rather, we are comparing the overall error of the six spectra. A spectrum with a smaller overall error is better than one with a larger error, no matter what the bias/variance ratio might be.

A detailed comparison showing the **average** errors of a **large** set of computed spectra using five different methods is given in reference [12]. The same conclusion, that quadratic windowing is better, is confirmed there.

4.7 Alternatives to Fourier Transform Analysis*

4.7.1 THE MAXIMUM ENTROPY METHOD

If we apply the spectrum estimation method to a process with a few sharp and closely spaced peaks, we would find that the peaks are poorly resolved as the window tends to merge fine details together. Yet, choosing a smaller bandwidth would not work because this increases variance. What we are up against is the basic weakness of the Fourier transform method, in that it assumes that any frequency is potentially of **equal importance**. When this is not true, the method cannot perform well. In such cases, the maximum entropy method tends to do better because it tends to reproduce the **most prominent** features of S(f). Less prominent structures, however, tend to be lost. Some care is therefore necessary when choosing this method since it must be applicable to the particular spectrum being estimated. Examples are found in underwater echo analysis where it has had some success [17].

The basic idea of the method is as follows: given x(t), we estimate R(t) from it, and fit an auto-regressive process to the auto-correlation by solving equations (2.77) and (2.78) for $k = 1, 2, \ldots, m-1$, producing the auto-regressive coefficient, $\alpha_k, k = 1, 2, \ldots, m$. We can then compute $<| X(f)|^2>$ from α as follows

Fourier transform both sides of equation (2.74) to give

$$X(f) = \sum_{i=1}^{m} \alpha_i \int \exp(j\omega t) x(t-i\Delta) dt + N(f) \qquad (4.51)$$

put $t = t' + i\Delta$. We then have

$$X(f) = \sum_{i=1}^{m} \alpha_i \exp(j\omega i\Delta) \int \exp(j\omega t') x(t') dt + N(f)$$

$$= \sum_{i=1}^{m} \alpha_i \exp(j\omega i\Delta) X(f) + N(f) \qquad (4.52)$$

or

$$X(f) = \frac{N(f)}{1 = \sum_{i=1}^{m} \alpha_i \exp(j\omega i\Delta)} \qquad (4.53)$$

so that

$$S(f) = <|X(f)|^2> = \frac{<|N(f)|^2>}{|1 - \sum_{i=1}^{m} \alpha_i \exp(j\omega i\Delta)|^2} \qquad (4.54)$$

The power spectrum of white noise, $<|N(f)|^2>$, is constant because it is the Fourier transform of a δ-function, which is the auto-correlation of white noise. Thus, $<|N(f)|^2> = N$ and we have

$$S(f) = \frac{N}{|1 - \sum_{i=1}^{m} \alpha_i \exp(j\omega i\Delta)|^2} \qquad (4.55)$$

Thus, once we know α, N, and the sampling interval Δ, we can easily find S(f) for x(t).

The correlation estimates, which provide a starting point for the maximum entropy method, can be derived from the signal series using one of the methods described in the next chapter. Burgh [18] gives an alternative approach which determines the required filter coefficients and driving noise power directly from the signal data and which does not depend on first obtaining an auto-correlation function.

A fast and simple recursive procedure for maximum entropy spectral analysis has been suggested by Anderson [19] which is capable of being programmed in a high-level language.

4.7.2 PISARENKO SPECTRAL DECOMPOSITION

The resolution of the maximum entropy method becomes poorer as the signal-to-noise ratio decreases [17]. An alternative modelling technique for spectral analysis which avoids this disadvantage has been suggested by Pisarenko [20]. This is known as Pisarenko decomposition and is found to alleviate the problem of degraded resolution of sinusoids at low signal-to-noise ratios.

The method is illustrated in Fig. 4.11 which shows that any signal can, in principle, be modelled by the addition of a number of weighted sinusoidal functions, together with white noise. Now the auto-correlation function of n sinusoidal functions plus white noise is equal to

$$R(k) = \sigma_n^2 \delta(k) + \sum_{i=1}^{n} \frac{Ai^2}{2} \cos(2\pi f_i k) \qquad (4.56)$$

$$0 \leqslant k \leqslant m$$

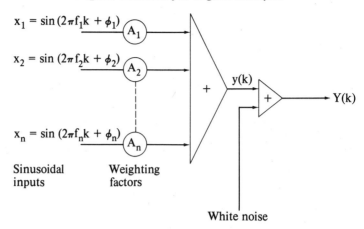

Fig. 4.11   Principle of the Pisarenko method

Here the white noise only affects the zero lag ($\tau = 0$) auto-correlation value. If the number of known non-zero lag values, m, is greater than or equal to twice the number of sinusoidal functions ($m \geq 2n$), the noise power can be determined exactly and subtracted from the auto-correlation zero lag value giving the signal auto-correlation function as

$$R_s(k) = R(k) - \sigma_n^2 \delta(k)$$

$$= \sum_{i=1}^{n} \frac{A_i}{2} \cos (2\pi f_i k) \qquad (4.57)$$

$$0 \leq k \leq m$$

Note that this is a completely noise-free auto-correlation function so that when this result is applied to the maximum entropy method it can yield infinitely fine two-sinusoidal resolution. Figure 4.12 shows a comparison of the resolution obtained with a three-sinusoidal signal for (*a*) Fourier, (*b*) maximum entropy and (*c*) Pisarenko methods (after Frost [17]).

4.7.3   WALSH SPECTRAL ANALYSIS

Representation of a time-series in the sequency domain gives a result that is not invariant to circular time shift as found with Fourier transformation. Fortunately the invariance is less important where the sum of the squares of pairs of transformed coefficients of the same sequency are taken and we can define the Walsh power spectrum as follows.

A sequency spectrum is obtained from the sum of the squares of the CAL and SAL function coefficients of the same sequency for each spectral

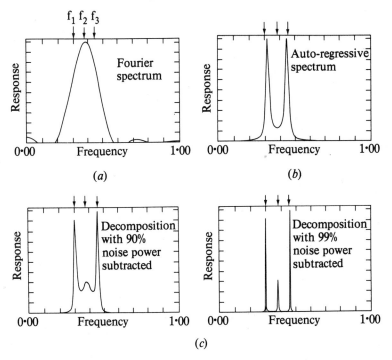

Fig. 4.12 Comparison of resolution obtained for (*a*) Fourier, (*b*) maximum entropy and (*c*) Pisarenko methods (after Frost)

coefficient in a manner similar to that used in Fourier analysis. Reference
to Fig. 3.14 and equation (3.98) will show that $N/2 + 1$ spectral points will
be obtained with the first and the last consisting of the square of the CAL
and SAL function only, viz.

$$
\left.
\begin{aligned}
P_{(0)} &= X_C^2(0) \\
P_{(k)} &= X_C^2(k) + X_S^2(k) \\
P\left(\frac{N}{2}\right) &= X_S^2\left[\frac{N}{2}\right] \\
k &= 1, 2, \ldots, \left(\frac{N}{2}\right) - 1
\end{aligned}
\right\}
\qquad (4.58)
$$

We can regard the Walsh spectrum so obtained as having a minimum
analysis bandwidth analogous to 'two degrees of freedom' obtained in the
Fourier case. A wider bandwidth is obtained by summing the squares of a
number of CAL and SAL coefficients for consecutive sequencies and

normalising by the number of pairs taken. Some examples of Walsh spectra are shown compared with the equivalent Fourier spectrum in Figs. 4.13 and 4.14.

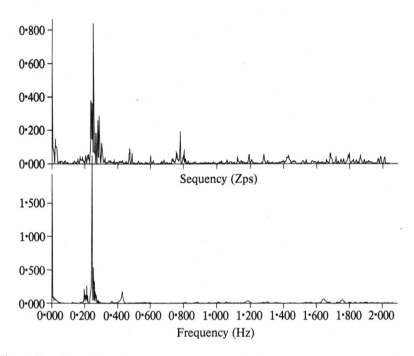

Fig. 4.13   Comparison between a sequency and frequency power spectrum for a
short-term transient signal
(*included by courtesy of Academic Press Ltd.*)

Figure 4.13 shows a comparison between Fourier and Walsh spectral analysis for a short-term transient signal indicating close similarities for the main regions of spectral power. However, a significant region of high sequency is present in the Walsh case, well removed from the main region found in both representations. This is due to high-order sequency components arising when a sinusoidally based signal is analysed in this way. A similar region of high frequency power is found with Fourier analysis of a rectangular synthesised signal, such as the comparative spectra of the pulse code modulation (PCM) wave-form shown in Fig. 4.14. The Walsh spectral representation of the PCM signal shows a precise sequency-limited bandwidth, due to the finite number of terms necessary to synthesise such a signal having a length binary-related to the Walsh time-base.

These examples indicate clearly the respective roles of Walsh and Fourier spectral analysis for discontinuous and smooth-varying signals.

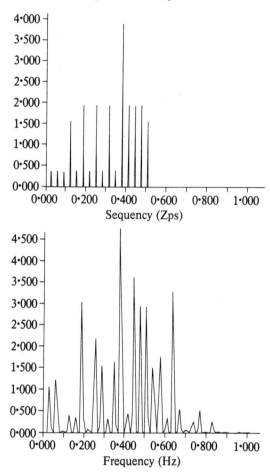

Fig. 4.14   Comparison between the sequency and frequency spectrum of a PCM
signal
(included by courtesy of Academic Press Ltd.)

Where the signal has a large bandwidth, however, e.g. on-line image
processing, the speed of obtaining the Walsh transform by digital computation may over-ride the performance consideration.

4.8   Interpretation

Processing of random signals is carried out generally for one of two
reasons: either we hope to improve the quality of the signal (e.g. removal of
unwanted elements, such as noise) or to interpret the processed signal in
order to gain insight into the mechanism which originally produced it.

Unfortunately the methods available for interpretation are not highly satisfactory and much is left to the experience and *a priori* knowledge of the investigator. This is particularly the case for spectrum interpretation where known methods of analysis having widespread applicability are limited to the derivation of only a few characteristics.

The simplest possible analysis one can perform of a computed spectrum is to compare it with a standard spectrum (template), familiar to anyone who has studied chemistry. When we wish to know if some substance is present in a given sample, we measure the radiation spectrum of the sample over some frequency range in which the spectrum of the substance is known, and compare the two. When the given substance is known to be a mixture, its spectrum would be some combination of the spectra of its constituents. But even then we may be able to find some frequency range over which the radiation emitted by the substance to be identified is dominant. We discuss the more difficult situation below.

Another simple spectrum analysis is the identification of sharp peaks. For example, in wind tunnel tests of aeroplane models, peaks in the vibration spectra of various parts of the model indicate the presence of resonances, which could lead to structural failing, and hence must be eliminated by design modification. Similarly, the spectrum of road and airport runway surfaces, after trend removal, should not contain sharp peaks, which indicate periodic structures that can cause severe vibrations in the vehicles travelling over them. Peaks in the spectrum of earthquake waves give information about the vibration modes, hence the structure of the earth. For electric signals, sharp peaks in spectrum always show feedback paths in the producing system. Thus, if the input signal is a sinusoid of frequency f, and the system is such that every peak/trough in the output is reinforced by the feedback from the previous peak/trough, then the oscillation is greatly magnified. Similar effects are caused by negative feedback if the feedback from every peak goes to a trough, making it deeper, and that from each trough goes to a peak. There are many other interpretations of sharp peaks in a spectrum depending on the signal source. Of course, in all cases we must be sure that the peaks are genuine, and are not caused by statistical fluctuations.

We have only discussed a few cases where simple, intuitive interpretation of results can be carried out. In many circumstances interpretation may involve very mathematical procedures, of which two examples are given in this book. One is composition analysis by least-squares fitting, Sub-section 4.8.1, and the other is system identification involving a noisy signal, Section 5.10.

4.8.1 COMPOSITION ANALYSIS

Let us consider the determination of the composition of a mixture by the analysis of its spectrum. Assume that the given sample contains constituents $1, 2, \ldots, m$, which have respective spectra $S_1(f), S_2(f), \ldots, S_m(f)$, and the mixture has the composite spectrum $S(f)$. Given the pre-condition that the different constituents emit radiation independently of each other, we have that $S(f)$ is just the sum of the individual spectra, multiplied by the **amount** of each substance. In other words

$$S(f) = \sum_{k=1}^{m} S_k(f)\delta_k \qquad (4.59)$$

where δ_k is the percentage content of substance k. δ_k may be 0 but not negative. (One must emphasise that the equation is not valid if the substances do not emit independently, for example, if radiation from material one may be absorbed by two, or if material three can cause induced emission from material four. The whole method breaks down then.)

The equation (4.59) cannot usually be solved exactly. Instead, we try to satisfy it approximately in the least-squares sense. That is, we seek δ_k's that will minimise the integrated square of error

$$\epsilon = \int_{-\infty}^{\infty} \left\{ S(f) - \sum_{k=1}^{m} S_k(f)\delta_k \right\}^2 df \qquad (4.60)$$

Elementary calculus provides that we find δ_k by taking the partial derivative of expression (4.60) and equating it to 0

$$0 = \frac{\partial\epsilon}{\partial\delta_k} = 2 \int_{-\infty}^{\infty} \left\{ S(f) - \sum_{k'=1}^{m} S_{k'}(f)\delta_k \right\} S_k(f) df \qquad (4.61)$$

Taking half of the terms to the other side we get the equation

$$\sum_{k'=1}^{m} \delta_{k'} \int S_k(f) S_{k'}(f) df = \int S(f) S_k(f) df \qquad (4.62)$$

What we have is just m linear equations in m unknowns $k, k' = 1, 2, \ldots, m$. In matrix form this is

$$\mathbf{A}\delta = \mathbf{B}$$

where $\mathbf{A}$ is an m x m square matrix, with

$$A_{kk'} = \int S_k(f) S_{k'}(f) df$$

and $\mathbf{B}$ is an m x 1 column vector, with

$$B_k = \int S_k(f) S(f) df$$

In actual computation we would know $S(f)$ and $S_k(f)$ over a finite set of points only, say M points. That is, we would know S at $S(f_i)$, $i = 1, 2, \ldots, M$, and we must replace integration by summation, viz.

$$
\left.
\begin{aligned}
A_{kk'} &= \sum_{i=1}^{M} S_k(f_i)S_{k'}(f_i), \quad k, k' = 1, 2, \ldots, m \\
\text{and} \\
B_k &= \sum_{i=1}^{M} S_k(f_i)S(f_i), \quad k = 1, 2, \ldots, m
\end{aligned}
\right\} \tag{4.63}
$$

However, the form of equation (4.62) is retained, and we still have m equations in m unknowns.

The above composition analysis procedure is simple in theory, but there is much potential hazard. First, the sample may contain substances not included among the m assumed constituents. Second, we might have included the spectrum of some assumed constituent which is not actually present. Both conditions can produce significant errors in the computed δ's, especially those of small magnitude. To guard against such possibilities, one should always examine the residuals,

$$
S(f_i) - \sum_{k=1}^{m} S_k(f_i)\delta_k \tag{4.64}
$$

If some of the values are quite large, then it is highly likely that the set of assumed constituents do not match the set actually present. There are what one calls **robust** estimation methods, which are less sensitive to such errors, and weighted least-squares methods, in which we give points with large errors smaller weights in a continuing refinement process. These are, however, beyond the scope of this book. Interested readers should consult references [21] and [22]. Reference [2] also discusses the problem of ill-conditioning, which is a third potential problem in composition analysis, and in least-squares approximation in general.

References

1. BINGHAM, C., GODFREY, M. D., and TUKEY, J. W. Modern techniques of power spectrum estimation. *IEEE Trans. Audio Electroacou.*, **AU-15, 2**, 56–66, 1967.
2. WIENER, N. *Extrapolation, Interpolation and Smoothing of Stationary Time Series.* MIT Press, Cambridge, Mass., 1949.
3. KHINTCHINE, A. Y. Korrelationstheorie der stationären stockastischen Prozesse. *Math. Annal.*, **109**, 604, 1934.

4. BLACKMAN, R. B., and TUKEY, J. M. *The Measurement of Power Spectra*. Dover, New York, 1959.
5. RABINER, L. R., and GOLD, B. *Theory and Applications of Digital Signal Processing*. Chapter 10. Prentice-Hall 1975.
6. PARZEN, E. Mathematical considerations in the estimation of spectra. *Technometrics*, **3**, 67–190, 1961.
7. JENKINS, G. M., and WATTS, D. G. *Spectral Analysis and its Applications*. Holden-Day, San Francisco, 1968.
8. FISHER, A. R., and YATES, F. *Statistical Tables*. Oliver and Boyd, London, 1967.
9. WONNACOTT, T. H. Spectral analysis combining a Bartlett window with an associated inner window. *Technometrics*, **3**, 235–43, 1961.
10. JENKINS, G. M. General considerations in the analysis of spectra. *Technometrics*, **3**, 133, 1961.
11. SLOAN, E. A. A comparison of linearly and quadratically modified spectral estimates of Gaussian signals. *IEEE Trans. Audio and Electroacoust.*, **AU-15**, 56–65, 1967.
12. YUEN, C. K. A comparison of five methods for computing the power spectrum of a random process using data segmentation. *Proc. IEEE*, **65**, 1977.
13. DROUILHET, P. R., and GOODMAN, L. M. Pole shared linear-phase band-pass filter bank. *Proc. IEEE*, **54**, 4 April 1966.
14. LERNER, R. M. Bandpass filters with linear phase. *Proc. IEEE*, **52**, 249–68, 1964.
15. ENOCHSON, L. D., and OTNES, R. E. An algorithm for digital one-third octave analysis. *J. Sound Vib.*, **2**, 4, April 1968.
16. BRENT, R. P., Algorithm A488. *Comm. ACM*, **17**, 704–6, 1974.
17. FROST, O. L. Power spectrum estimation. *N.A.T.O. Advanced Study Institute on Signal Processing*, Portovenere, Italy, August/September 1976.
18. BURGH, J. P. *Maximum Entropy Analysis*. Ph.D. Thesis, Stanford University, Stanford, California, U.S.A. 1973.
19. ANDERSON, N. On the calculation of filter coefficients for maximum entropy spectral analysis. *Geophysics*, **39**, 69–74, 1974.
20. PISARENKO, V. F. The retrieval of harmonics from a covariance function. *Geophysics J. R. Astron. Soc.*, 347–366, 1973.
21. LAWSON, C. L., and HANSON, R. J. *Solving least-square problems*. Prentice-Hall, Englewood Cliffs, N.J., 1974.
22. BLACKBURN, J. A. *Spectrum analysis*. Dekker, New York, 1970.

Chapter 5

Correlation Analysis

5.1 Introduction

This chapter is concerned with those mathematical methods and computational techniques used to determine the similarity between events and to distinguish between signal and noise in a composite wave-form. Statisticians have long been using measurement techniques which establish the dependence of one set of numbers upon another. Historically the first ideas leading to a quantitative measure of similarity stem from this work, which was originally confined to real data. The development of electrical communications and the need to improve signal-to-noise ratios have led engineers and scientists to develop correlation methods applicable to complex data. These have been derived from their work on Fourier series and transforms and, in particular, methods of using these to transform from a real domain (e.g., time domain) into a complex (frequency) domain. The realisation that, under given conditions, power spectral density functions and correlation functions are Fourier transform pairs, gave considerable impetus to this work, with the result that correlation techniques have been accepted during the last two decades as a major scientific tool, applicable to many areas of research. It is interesting to note that within the last few years developments in electronic equipment and hybrid computation have stimulated considerable interest in correlation using time domain methods, giving fast alternative methods of correlation estimation.

Following a consideration of the mathematical basis for correlation a number of analysis techniques will be discussed and will include both transform and direct evaluation methods. Finally several important applications of correlation will be described including that of signal-to-noise enhancement.

5.2 Statistical Basis for Correlation

In Chapter 2 we defined the correlation coefficients of two random variables and said that it in some sense indicates how 'closely related' the two variables are. To be more quantitative, correlation is a measure of a linear relationship between two variables. Though we generally mean random variables, correlation is also applied to deterministic functions.

A basic indication of the correlation between two variables x and y is the sum of the products, Σxy. Consider the set of independent data measurements given in Table 5.1(*a*) and 5.1(*b*). x and y take on the same set of values, 0 to 7, but in (*a*) the values of x and y are less closely related than in (*b*). We notice that the sum of products is larger for (*b*). If we plot the values on a two-dimensional graph (Fig. 5.1), we see that the second set of data (circles) show a nearly linear relationship, while the first set (crosses) do not.

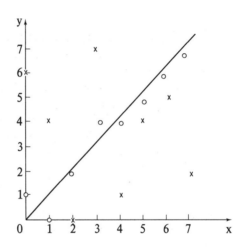

Fig. 5.1   Correlated and uncorrelated data sets

Table 5.1   **Relation between statistically independent samples**

(a)	x	y	xy	(b)	x	y	xy
	1	4	4		0	1	0
	2	0	0		1	0	1
	0	6	0		2	2	4
	7	2	14		3	4	12
	5	4	20		4	4	16
	3	7	21		5	5	25
	6	5	30		6	6	36
	4	1	4		7	7	49
			93				143

We recall that in Chapter 2 we defined correlation by ensemble averaging. In actual computation we must use time averaging. Given two series of values $x_1 - x_N$ and $y_1 - y_N$, and if $\bar{x}$ is the mean of x

$$\bar{x} = \frac{1}{N} \sum_{n=1}^{N} x_n \tag{5.1}$$

and $\bar{y}$ is defined similarly as the mean of y, then $\delta x_n = x_n - \bar{x}$ and $\delta x_n = y_n - \bar{y}$ are respectively the deviations from average. We then take another average of the products of the N deviations

$$C_{xy} = \frac{1}{N} \sum_{n=0}^{N} \delta x_n \delta y_n \tag{5.2}$$

This is the measured covariance between x and y. The correlation coefficient, ρxy, however, is a normalised version of the covariance, and is defined as the covariance divided by the product of the standard deviations of x and y, namely σx and σy,

$$\rho_{xy} = C_{xy}/\sigma_x \sigma_y \tag{5.3}$$

Note that we can compute σ_x by taking the square root of C_{xx}. As in Chapter 2, we have $-1 \leqslant \rho_{xy} \leqslant 1$. When $|\rho_{xy}| \simeq 1$ x and y are said to be highly correlated. If ρ_{xy} tends to be zero then the two variables are approximately independent.

5.3 Applicability to Signal Processing

In the application of correlation methods to signal processing problems, the features whose similarity is to be assessed are generally expressed in sampled values. For the following theoretical derivations, however, we shall denote the variables as continuous functions. Also, the actual variables may be functions of **space** rather than **time**, e.g., a picture is a two-dimensional spatial function. However, in the following we shall proceed as if we are handling only functions of time. The generalisation to other cases is not difficult, as shown in Chapter 7.

Given signals $x(t)$ and $y(t)$, we have the time average

$$\bar{x} = \frac{1}{T} \int_0^T x(t)dt \tag{5.4}$$

and similarly for y. The deviation from the mean is

$$\delta x(t) = x(t) - \bar{x} \tag{5.5}$$

with an analogous expression for $\delta y(t)$. In a practical covariance/correlation estimation problem the similarity between x and y may become apparent only after a time delay. For example, the similarity between the input (stimulus) and output (response) signals of a system would be delayed by the response lag of the system, say τ. One would expect $y(t+\tau)$ to be highly correlated with $x(t)$. Since, however, the system delay would in general not be known beforehand, we are forced to correlate $y(t+\tau)$ with $x(t)$ for all possible values of τ in order to detect the similarity between x and y. Thus, we require the general covariance function

$$C_{xy}(\tau) = \frac{1}{T} \int_0^T [x(t)-\bar{x}][y(t+\tau)-\bar{y}]dt \qquad (5.6)$$

(It is necessary to assume that x and y are both time-invariant variables, so that $\bar{x}$ and $\bar{y}$ do not depend on t and C_{xy} is a function of the time difference τ only.) The correlation coefficient function corresponding to definition (5.3) is the normalised form of equation (5.6), being

$$\rho_{xy}(\tau) = C_{xy}(\tau)/(\sigma_x\sigma_y) = C_{xy}(\tau)/\{C_{xx}(0)C_{yy}(0)\}^{1/2} \qquad (5.7)$$

It is easy to see from equation (5.6) that

$$C_{xy}(\tau) = \frac{1}{T} \int_0^T x(t)y(t+\tau)dt - \bar{x}\bar{y} = R_{xy}(\tau) - \bar{x}\bar{y} \qquad (5.8)$$

Here $R_{xy}(\tau)$ is the measured cross-correlation between the two variables

$$R_{xy}(\tau) = \frac{1}{T} \int_0^T x(t)y(t+\tau)dt \simeq <x(t)y(t+\tau)> \qquad (5.9)$$

In analogy to equation (2.67) we have

$$\rho_{xy}(\tau) = \{R_{xy}(\tau)-\bar{x}\bar{y}\}/\{R_{xx}(0)-\bar{x}^2\}^{1/2}\{R_{yy}(0)-\bar{y}^2\}^{1/2}$$

Sometimes, a normalised form of equation (5.9) is employed

$$r_{xy}(\tau) = R_{xy}(\tau)/\{R_{xx}(0)R_{yy}(0)\}^{1/2} \qquad (5.10)$$

If x and y are zero-mean random processes then $R_{xy} = C_{xy}$, and in this case $r_{xy} = \rho_{xy}$. But otherwise the two should not be confused.

5.4   Auto-Correlation

In the previous section we were concerned with the relation between two random processes. The purpose of auto-correlation is to analyse the relation between the values of the same random process at different times. This was first introduced in Chapter 2, where we showed the auto-correlation of several simple random processes. In the last chapter we discussed its relation to the power spectrum. We are now ready to start more detailed discussion. By definition

$$R_x(\tau) = <x(t)x(t+\tau)>$$ (5.11)

If the signal is ergodic we can approximate ensemble averages by time averages, so that

$$R_x(\tau) = \frac{1}{T} \int_0^T x(t)x(t+\tau)dt$$ (5.12)

This corresponds to equation (5.9) for x = y. Equation (5.8) would become

$$C_x(\tau) = R_x(\tau) - \bar{x}^2$$ (5.13)

If x is a zero-mean process, or, if we make it one by subtracting $\bar{x}$ from each value of x, then equation (5.12) would produce $C_x(\tau)$. Since the condition of zero-mean can always be made to hold, we may assume that all processes we encounter have zero-mean, though we may need to carry out pre-processing in order to arrive at this condition.

To illustrate the discussion, Fig. 5.2 shows the theoretical auto-correlation of sinusoidal wave-form and that of a square wave-form, taken

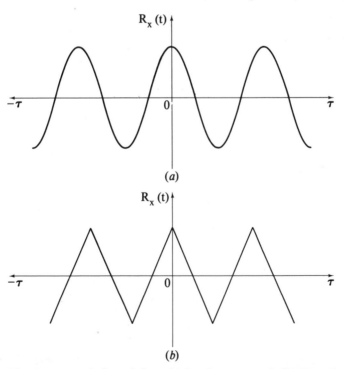

(a)

(b)

Fig. 5.2   Auto-correlation of sinusoidal and square periodic wave-forms

over infinite intervals, i.e. Fig. 5.2(*a*) shows, for x = A sin (ωt+φ)

$$R_x(\tau) = \lim_{T\to\infty} \frac{A^2}{2T} \int_{-T}^{T} \sin(\omega t+\phi)\sin[\omega(t+\tau)+\phi]dt \qquad (5.14)$$

By expanding the sines using trigonometric identities we find that all terms but one would vanish, leaving

$$R_x(\tau) = {}^1\!/_2 A^2 \cos \omega\tau \qquad (5.15)$$

Thus, the auto-correlation of a sinusoid is a cosine function. Note that the maximum value of R is equal to the mean-square value of x (or half of its peak squared value).

For both plots in Fig. 5.2 x(t) is periodic. Their auto-correlations have the same period. We also see that R is an even function, i.e.

$$R_x(\tau) = R_x(-\tau) \qquad (5.16)$$

Thus, we could equally well write

$$R_x(\tau) = <x(t)x(t-\tau)> \qquad (5.17)$$

rather than equation (5.11). We might consider equation (5.11) as correlating x with its own future, while equation (5.17) correlates x with its past.

Other properties of R_x are

$$R_x(0) = \frac{1}{T} \int_0^T x^2(t)dt = \overline{x^2(t)} \qquad (5.18)$$

and that

$$R_x(0) \geq R_x(\tau) \qquad (5.19)$$

To demonstrate the latter, we consider the integral

$$\frac{1}{T} \int_0^T [x(t)\pm x(t+\tau)]^2 dt$$

This cannot be negative. Expanding out the square term we get

$$\frac{1}{T} \int_0^T x^2(t)dt + \frac{1}{T} \int_0^T x^2(t+\tau)dt \pm \frac{2}{T} \int_0^T x(t)x(t+\tau)dt$$

Applying equation (5.12), we have $R_x(0) + R_x(0) \pm 2R_x(\tau)$. As the whole expression is never less than zero this gives

$$2R_x(0) \geq \pm 2R_x(\tau)$$

These results, derived using time averages, have their counterparts derived using ensemble averages in Chapter 2.

Using equations (5.12) and (5.18) we can restate the normalised version of the auto-correlation function as

$$r_x(\tau) = R_x(\tau)/R_x(0) = \int_0^T x(t)x(t+\tau)dt / \int_0^T x^2(t)dt \qquad (5.20)$$

The particular importance of the auto-correlation function lies in the fact that, in addition to providing information about the behaviour of x(t) in the time domain, it also, indirectly, tells us about the frequency domain behaviour of x, as provided by the Wiener theorem of Section 4.2.2. In fact, by inspecting the plot of $R_x(\tau)$ against τ, we can get a very good idea about the type of signal x is. In Fig. 5.3(*a*) we show the auto-correlation of a wide-band noise wave-form. (This is similar to the white noise introduced in Chapter 2.) The auto-correlation is nearly zero except for a sharp peak at $\tau = 0$, showing that x(t) at different times are nearly unrelated with each other, and that only a small time shift destroys the similarity between x(t) and x(t+τ). In contrast, a sinusoidal signal, as shown in Fig. 5.2(*a*), has a correlation function which persists indefinitely over an infinite range of time lags, and there is a persistent similarity between different values of x(t) regardless of how large the time lag might be. (The reason is of course that

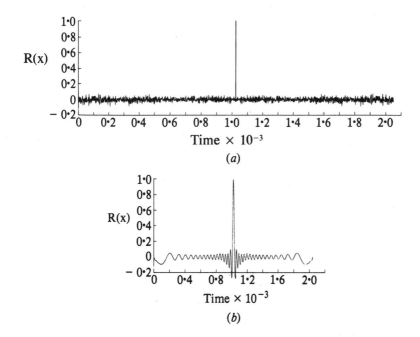

Fig. 5.3   (*a*) Auto-correlogram of a wide-band noise wave-form
             (*b*) Auto-correlogram of a narrow-band noise wave-form

x(t) is in this case periodic, so that it repeats itself indefinitely.) We may also note that the rate of decay of R_x gives a good indication of the bandwidth of x. A 'narrow band' signal like a sinusoid, which has only one frequency value, has a persistent auto-correlation function, while a broad band noise has an $R_x(\tau)$ that goes rapidly to zero when τ goes away from the origin.

In fact, Wainstein and Zubakov [1] have derived an 'uncertainty relation' stating that the length of time over which R_x is significant, $\delta\tau$, is related to the bandwidth of the process, $\delta\omega$, by

$$\delta\omega . \delta\tau \simeq n/2 \qquad (5.21)$$

If the wave-form has a random plus a periodic component, the auto-correlation function will take on a form similar to that given in Fig. 5.3(*b*). The plot exhibits the periodicity of the repetitive signal whilst reducing the contribution of the additive noise as τ increases to larger delay times. We see that the auto-correlation function provides a powerful means of identifying periodicities hidden in random noise.

The time average definition given in equation (5.12) breaks down when we have a signal defined on the interval [0,T] only, because x(t+τ) is undefined for any t $>$ T$-\tau$, and the integration cannot be carried out. (The difficulty does not arise if x is periodic, as in such cases x can be extended periodically to any value of t+τ.) In fact, the upper limit of integration can only be T$-\tau$, not T. This then brings in the question of whether we should replace the scaling factor $(T)^{-1}$ by $(T-\tau)^{-1}$. This is in fact a particular example of bias/variance trade-off, and is discussed in a subsequent section.

5.5   Cross-Correlation

Despite their apparent mathematical similarity there are several important differences between the auto- and cross-correlations. While $R_x(\tau)$ is always symmetric about the origin and has a maximum value there, the structure of $R_{xy}(\tau)$ is not necessarily so. In fact

$$R_{xy}(\tau) = <x(t)y(t+\tau)> = <y(t)x(t-\tau)> = R_{yx}(-\tau) \qquad (5.22)$$

(The same formula is obtained if we use time rather than ensemble averages.) There is no general relation between $R_{xy}(\tau)$ and $R_{xy}(-\tau)$ itself. Also, we have as special cases of the averaging relationships given by equations (2.23) and (2.24) the following relation

$$| R_{xy}(\tau)|^2 \leq R_x(0)R_y(0) \qquad (5.23)$$

and also

$$| R_{xy}(\tau)| \leq \frac{1}{2}[R_x(0)+R_y(0)] \qquad (5.24)$$

Thus, the values of R_{xy} are bounded by the maximum values of R_x and R_y, but there is no limiting relation between its own values. The inequalities are useful in actual computation for checking the correctness of results, especially as $R_x(0)$ and $R_y(0)$ are just the mean-squares of x and y and so can be found easily.

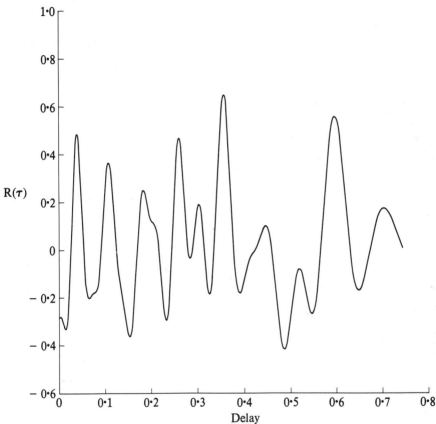

Fig. 5.4   A cross-correlogram

A typical computed cross-correlation function is shown in Fig. 5.4. As we see, its structure is quite different from that of the auto-correlations of Fig. 5.2. The fact that $R_{xy}(\tau)$ does not equal $R_{xy}(-\tau)$ actually means that the cross-correlation contains information absent in auto-correlation, namely, information about **phase differences**. This will be explained below.

Suppose we cross-correlate $x = A \sin(\omega t + \phi)$ with $y = B \sin(\omega t + \phi')$. That is, we compute the infinite time average

$$R_{xy}(\tau) = \lim_{T \to \infty} \frac{AB}{2T} \int_{-T}^{T} \sin(\omega t + \phi)\sin[\omega(t+\tau) + \phi']dt \qquad (5.25)$$

This produces a result similar to equation (5.15)

$$R_{xy}(\tau) = \tfrac{1}{2}AB \, \cos(\omega\tau + \phi - \phi') \tag{5.26}$$

That is to say, when two sinusoids of the same frequency are correlated the result is a sinusoid with that frequency and a phase which is the **phase difference** of the two functions correlated. Also, the magnitude of the cross-correlation increases with the amplitude of either function. Cross-correlation, then, provides a method for extracting information about a periodic signal buried in extraneous noise: we can find its amplitude and phase if we correlate the mixture with a sinusoid of the same frequency. But since any function can be considered as a linear combination of sinusoids (Fourier series), we can in fact extract information about arbitrary functions in this way. As we saw in the previous chapter, the Fourier transform of the cross-correlation is the cross-spectrum $S_{xy}(f)$, which provides information about the amplitude and phase relations of $X(f)$ and $Y(f)$. When $X(f)$ is known, we then learn something about $Y(f)$, and hence $y(t)$. Occasionally, however, the method fails. For example, if we correlated $x = A \sin(\omega t + a) + B$ with $y = C \sin(\omega t + c) + D \sin(n\omega t + d)$, the result is in fact

$$R_{xy}(\tau) = \tfrac{1}{2}AC \, \cos(\omega\tau + a - c) \tag{5.27}$$

The B and D terms have been lost, and information about y is only partially recovered. Consequently, one needs to apply a variety of analysis to the same signal to gain a full understanding.

5.6   Errors in Auto- and Cross-Correlations

We explained in Chapter 2 that time and ensemble averaging produces equal results for ergodic processes. In actual computation we cannot perform summation or integration over infinite time, and the measured averages are themselves random quantities. As a result, they depart from their theoretical values. As correlations are, by definition, average quantities, they suffer from the same errors of estimation that we discussed in Section 2.5, namely bias and variance. It is possible to carry out a detailed error analysis as was done in Section 2.5.2 for μ and σ. This, however, would be rather lengthy. We shall instead present a more restricted analysis, which makes the point quite effectively.

Consider again $x = \sin(\omega t + \phi)$. Let us now suppose that we have measured its values over the interval [0,T] only. To estimate $R_x(\tau)$, we can only integrate $x(t)x(t+\tau)$ between 0 and $T-\tau$, as otherwise we would require $x(t)$ for $t<0$ or $t>T$, which is not available. It is natural to replace equation (5.12) by

$$R_x(\tau) = \frac{1}{T-\tau} \int_0^{T-\tau} x(t)x(t+\tau)dt \tag{5.28}$$

We have then

$$R_x(\tau) = \tfrac{1}{2}A^2\cos\,\omega\tau - \frac{A^2\cos[\omega(T-\tau)]}{4\omega(T-\tau)}\;\sin(\omega\tau+2\phi) \qquad (5.29)$$

The first term is recognised as the theoretical auto-correlation, and the other is an error term. For $\tau\sim0$ the error should be small, but if $\tau\sim T$ then $T-\tau\sim0$ and the second term would be very large. Consequently, it pays to reduce the size of the second term, even at the expense of making the first term somewhat less accurate. This is why we should in fact approximate R_x, not by equation (5.28), but by

$$R_x(\tau) = \frac{1}{T}\int_0^{T-\tau} x(t)x(t+\tau)dt \qquad (5.30)$$

This is not only simpler it also turns out to be more accurate on average, as we now have

$$R_x(\tau) = \frac{T-\tau}{2T}A^2\cos\omega\tau - \frac{A^2\cos[\omega(T-\tau)]}{4\omega T}\;\sin(\omega\tau+2\phi) \qquad (5.31)$$

For small τ this is somewhat less accurate than equation (5.29), as both terms contribute towards the error. But for τ close to T there is much less error because the second term is now significantly smaller. We thus see that the more 'natural' expression (5.28) is in fact not as good as expression (5.30), in which we have greatly reduced the term corresponding to variance at the cost of introducing a small bias in the first term. Note also that the length of the interval, T, must be such that $1/\omega T$ is small as the error is directly related to this factor. In general, the error term is related to $1/BT$, where B is the bandwidth of the random process. This holds for both auto- and cross-correlations. (Again we see the importance of the product BT, as found in Chapter 4.)

5.7 Digital Methods

The general problem in digital computation of correlation functions is one of speed and efficiency, as well as that of program size and the ability to process long data records. We shall discuss a number of computing techniques here, starting with some general remarks. First, a discrete form of equation (5.9) is

$$R_{xy}(\tau) = \frac{1}{N-\tau}\sum_{i=1}^{N-1} x_i y_{i+\tau}, \quad \tau = 0,1,\ldots,m \qquad (5.32)$$

where m is the total number of correlations we need, and N is the number of data points. Equation (5.32) is an unbiased estimate of the ensemble average definition (5.9) for a finite time period, and shows that the summation period will decrease as the correlation time lag, τ, is made larger. We would therefore expect the reliability of this estimate to be reduced as the lag approaches the record length N. As explained in the previous section, for $\tau < N$ we can, with little loss in accuracy, replace equation (5.32) by

$$R_{xy}(\tau) = \frac{1}{N} \sum_{i=1}^{N-\tau} x_i y_{i+\tau} \qquad (5.33)$$

In fact, the accuracy is **improved** for $\tau \simeq N$, though in practice we do not compute R for large τ's, so that this point seldom arises.

The direct evaluation of equation (5.33) on a digital machine involves large amounts of computation time, as approximately N^2 multiplications would be required to evaluate all R. However, we saw in Chapter 3 that convolution (identical to correlation but for a sign change for τ) can be obtained by carrying out two forward Fourier transformations, one inverse transformation, and one multiplication per data sample pair. If the FFT algorithm is taken advantage of fully in the computation we would only need $3N.\log_2 N + N$ multiplications. This is considerably faster than N^2 multiplications for large N. (Note however that in the FFT method we need complex arithmetic. Arithmetic is real for equation (5.33).) For long data vectors, therefore, the FFT method is now commonly used. However, there are also several simpler techniques for reducing calculation time that can be applied in special problems, and these will also be considered below. Some existing methods work by exchanging multiplications for additions using a grouping arrangement [2]. Other techniques convert products into sums of squares, which can be computed by table look-up. Another important class of methods is based on the reduction in the quantisation levels that the sample data can take.

5.7.1 THE HALF-SQUARES METHOD

This was suggested by Schmid [3], based on the observation that

$$xy = \tfrac{1}{2}[(x+y)^2 - x^2 - y^2] \qquad (5.34)$$

so that equation (5.33) can be expressed as

$$R_{xy}(\tau) = \frac{1}{2N} \left[\sum_{i=1}^{N-\tau} (x_i + y_{i+\tau})^2 - \sum_{i=1}^{N-\tau} (x_i)^2 - \sum_{i=1}^{N-\tau} (y_{i+\tau})^2 \right] \quad (5.35)$$

Assuming that we provide a function table which contains all possible values of x^2, y^2 and $(x+y)^2$, we can then evaluate equation (5.35) by table look-ups and additions. If x, y and x+y are no more than 8 bits in length, say, then there are only 2^8 different values of squares, and the table is only 256 words long. In comparison, if we store all possible values of xy, the table would be 2^{16} words long, though only one table look-up would be needed for each product. Thus, the half-squares method trades computing time for storage. A similar method, called the quarter-squares method, replaces equation (5.34) by

$$xy = \frac{1}{4}[(x+y)^2-(x-y)^2] \tag{5.36}$$

5.7.2 COARSE QUANTISATION

The computation process (5.33) is one that greatly reduces round-off errors because it adds a large set of numbers and divides the sum by N. Consequently one can use very inaccurate data but still produce fairly good results. We can thus quantise x and y coarsely, i.e., give x only a small number of bits. For a given class of random processes, coarse quantisation will permit considerable saving in digital equipment, particularly in the analog-to-digital conversion and storage hardware. Experience has shown that as many as 5 bits could be saved from originally 13-bit data [4]. (Pictorial data often produce a poor appearance when coarsely quantised. This can be overcome partially by adding a broad band noise, called 'dither', to the signal before quantisation. The randomisation effect of the dither is sometimes desirable for other signals too.)

The most significant hardware simplification and reduction in computing is attained when we employ single bit quantisation, or **clipping**. Either one or both signals are quantised to only two levels, positive or negative, so that only the algebraic sign of the signal is preserved. It can be shown that provided the signals are Gaussian random processes and that T is made long enough, the theoretical correlation function can be accurately reproduced [5]. Weinreb [6] has shown that one should increase the length of the data by a factor of 2·5 to achieve the same accuracy attainable with full length quantised data.

5.7.3 CORRELATION USING THE FFT

In the last chapter we saw that the power spectrum is the Fourier transform of the auto-correlation, and the cross-spectrum is the transform of the cross-correlation. For some problems correlations are easier to compute than spectra, and we would compute the latter from the former. But we may also encounter the reverse situation, which is in fact more frequent. The same theorem then permits us to compute correlations from spectra.

While the underlying theory remains unchanged, the practical details are now rather different. We saw that it is always necessary to perform a great deal of averaging or smoothing operations on the spectrum to obtain satisfactory results. This must *not* be done if we wish to compute the correlation. As we showed in Section 4.2.2 smoothing the spectrum is equivalent to multiplying a weighting function into the correlation. Now whereas smoothing the spectrum is beneficial, in that the resulting estimate would have less leakage and variance and hence provide a better approximation to the theoretical spectrum, there is no corresponding beneficial effect as far as the correlation is concerned, since the mathematical operation carried out is one of multiplication. (This is just an example of a general phenomenon, that a small error in the frequency domain may correspond to a large error in the time domain, and vice versa.) To compute correlations, then, we should take the Fourier transform of the **raw** spectrum, i.e. $|X(f)|^2$ for R_x and $X(f)^*Y(f)$ for R_{xy}.

The FFT method is summarised as follows:

1. Add N zeros to each x and y series. Take a 2N-point FFT on each vector.
2. Multiply each Y into X*. (If we are computing the auto-correlation, x = y; then this produces X*X, or $|X|^2$.
3. Take a 2N-point inverse FFT on the vector of products.

Some remarks on the addition of zeros may be given here. It was stated in Section 4.5.1 that, whenever we compute $|X(f)|^2$ we must use the frequencies $f = i/2T$, not i/T, and that doubling the length of x(t) by adding zeros achieves this purpose. The same requirement holds here. If zeros are not added the computed correlation would in fact be a half-length compression of the correct result namely

$$R_x^c(\tau) = R_x(\tau) + R_x(N - \tau) \tag{5.37}$$

This is known as circular correlation and is illustrated in Figs. 5.5 and 5.6.

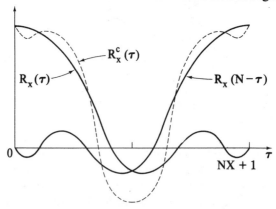

Fig. 5.5 Circular correlation

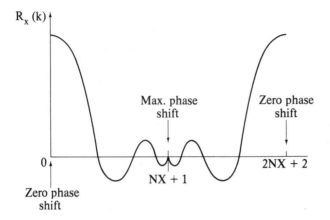

Fig. 5.6   Effect of adding zeros

In Fig. 5.5, the interval, T, is only half as long as required, and because R_x is symmetric, it and its mirror image co-exist in the interval, giving a combined result R_x^c. In Fig. 5.6 the interval is now doubled by zero insertions, and R_x is clearly separated from its own mirror image.

The R_x or R_{xy} produced in this method is the biased estimate of equation (5.33), not the unbiased estimate of equation (5.32). The bias can be removed simply by multiplying the results into $N/(N - \tau)$. Again, we believe this is actually unnecessary and generally the biased estimate is preferable.

5.8   Recovery of Signal from Noise

The acquisition of data relating to a given physical phenomenon is never obtained without acquiring some other, unwanted information. This generally includes unpredictable disturbances of various kinds, though periodic interference and transients are also known. (The latter are easier to remove than random disturbances.) The presence of this random element sets a limit to the attainable precision of measurement for the often small value of useful information contained within the acquired signal. Even assuming perfect data acquisition and processing equipment, certain random effects due to quantum mechanical behaviour may still occur.

Thus, we have the basic problem of improving the signal-to-noise ratio of the measured signal. The technique used of course depends on the particular situation. In radio communication, for example, the solution is filtering, i.e., the received signal is passed through highly selective circuits which resonate and so generate a large output signal in the region of the wanted frequency. Simple filtering like this is of little use, however, when signal and noise occupy the same part of the frequency spectrum. Here a statistical approach to the problem can produce valuable results for those cases where the signal is repetitive, permitting an averaging method to be applied. That is, if we add up a number of observations of a repeating signal, the contribution provided by the random noise will tend to cancel out, while the summation of the repetitive signal will result in increased output. A similar result can be derived for a **single** piece of signal by correlation methods, even when the signal is not periodic. This provides a powerful technique for extracting information about signals buried in noise.

Thus, in general we have three techniques for signal recovery. The filtering method will be considered in the next chapter, while the other two will be discussed below. The choice of method is dependent on the known characteristics of a signal and, for the case of recorded information, on its form of storage as well. In many situations a practical choice will be determined by the available equipment, but where a wider choice is possible, considerable economy in processing costs can be achieved. For example, where a real time environment is concerned, such as the continuous analog tape recording of low frequency oceanographic data, it would be unrealistic to digitise the mass of data before attempting a signal recovery operation. Instead, an analog data reduction operation would need to precede the digitisation of the reduced data, or alternatively, use be made of a real time hybrid system specifically designed for this purpose.

5.8.1 USE OF CORRELATION

The extent to which correlation contributes to the improvement of the signal-to-noise ratio has been described in a notable paper by Lee et al. [7]. The main observation of the paper was that the signal-to-noise ratio in the cross-correlation between the signal and the signal-to-noise mixture is considerably greater than the corresponding ratio in the mixture itself. That is, given signal $x(t)$ and noise $n(t)$, and the task being to extract information about the presence of $x(t)$ in the mixture $y(t) = x(t) + n(t)$, we consider the cross-correlation

$$R_{xy} = <x(t)[x(t+\tau) + n(t+\tau)]> = R_x(\tau) + R_{xn}(\tau) \qquad (5.38)$$

The signal-to-noise ratio in $y(t)$ is $R_x(0)/R_n(0)$, or $<x^2>/<n^2>$. We denote this value as $\frac{1}{r}$. Now consider the corresponding ratio in equation (5.38) for an actual piece of signal. The ensemble averages are now estimated using time averages. The desirable part of R_{xy}, which is R_x, is

$$R_1 = \frac{1}{T} \int_0^T x(t)x(t+\tau)dt$$

while the undesirable part is

$$R_2 = \frac{1}{T} \int_0^T x(t)n(t+\tau)dt$$

The average power of the former is

$$<[R_1(\tau)]^2> = \frac{1}{T^2} \iint <x(t)x(t+\tau)x(s)x(s+\tau)>dtds$$

Making use of equation (2.50) we get

$$<[R_1(\tau)]^2> = \frac{1}{T^2} \iint [<x(t)x(t+\tau)><x(s)x(s+\tau)>+<x(t)x(s)><x(t+\tau)$$
$$x(s+\tau)>+<x(t)x(s+\tau)><x(t+\tau)x(s)>]dtds$$

$$= \frac{1}{T^2} \iint \{[R_x(\tau)]^2+[R_x(s-t)]^2+R_x(s+\tau-t)R_x(s-\tau-t)\}dtds$$

$$\simeq [R_x(0)]^2$$

whereas $<[R_2(\tau)]^2> = \frac{1}{T^2} \iint <x(t)n(t+\tau)x(s)n(s+\tau)>dtds$

$$= \frac{1}{T^2} \iint \{[R_{xn}(\tau)]^2+R_x(s-t)R_n(s-t)+R_{xn}(s+\tau-t)R_{xn}(s-\tau-t)\}dtds$$

$$\simeq R_x(0)R_n(0)$$

Thus the percentage of power for signal in the computed R_{xy} is approximately

$$[R_x(0)]^2/\{[R_x(0)]^2+R_x(0)R_n(0)\} = [1+R_n(0)/R_x(0)]^{-1} = (1+r)^{-1} \tag{5.39}$$

Now suppose we estimate the auto-correlation of y. We have

$$R_{yy}(\tau) = <[x(t)+n(t)][x(t+\tau)+n(t+\tau)]>$$
$$= R_{xx}(\tau)+R_{nn}(\tau)+2R_{xn}(\tau)R_{nx}(\tau) \tag{5.40}$$

Carrying out a similar analysis as before we find that the desirable part of estimated R_{yy} has the average power $\simeq [R_x(0)]^2$, but the two undesirable parts would have mean power $\simeq [R_n(0)]^2$ and $\simeq 2R_x(0)R_n(0)$, so that the percentage of signal in the computed correlation is

$$[R_x(0)]^2/\{[R_x(0)]^2+[R_n(0)]^2+2R_x(0)R_n(0)\} = (1+2r+r^2)^{-1} \tag{5.41}$$

194   *Digital Methods for Signal Analysis*

Independently of the value of r, ratio (5.40) is more acceptable than that given by equation (5.41). Thus our discussion indicates that, analysing x(t) and y(t) together is more effective, by a factor of $(1+2r+r^2)/(1+r)$, than analysing y(t) by itself, as far as detecting the presence of x(t) in y(t) is concerned. Figure 5.7 illustrates this use of the reference wave correlation method. Here the noisy signal y(t) is correlated with the reference wave x(t) as well as with itself, and the respective correlations, or equivalently, the auto- and cross-spectrums, show the clear superiority of the latter method.

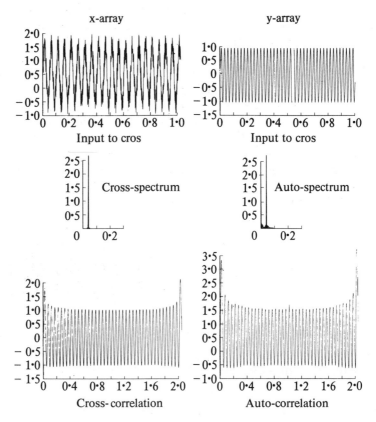

Fig. 5.7 Use of a reference wave-form to extract a periodic wave-form from a noisy signal

The cross-correlation method for signal extraction arose largely from the needs of radar signal processing, in which the presence, as well as the positions, of distant objects are determined by detecting the radar waves reflected by these objects. (Sonar works on a similar principle.) The ability to dctect the return signal reflected by a target can be shown to be directly proportional to the product of the average signal power and the signal duration. Practical limitations in transmitter design control the peak value of the signal and it becomes necessary to obtain the required range by increasing the signal duration. Unfortunately the reduction in bandwidth which this entails also leads to a reduction in signal resolution, i.e., the ability to distinguish between echoes derived from closely spaced objects. A solution to this resolution/range problem has been proposed by Wiener, Shannon and others. This suggests the replacement of the long narrow-band pulse by a swept-frequency signal having both long duration and wide bandwidth. This technique is referred to as pulse compression and contributes enormously to the range and resolution capability of modern radar equipment. Detection of the swept-frequency echo signal and the separation of this from its associated wave-form is carried out using cross-correlation of the signal with a reference wave-form derived from the transmitter output. This process compresses the energy dispersed within the long swept-frequency signal into a short well-defined pulse of consequently enhanced average power. This is shown in Fig. 5.8 and illustrates the value of this technique in raising the correlated pulse to a value well clear of the noise level. A 'threshold detection' can then be carried out to give a precise location of the time position of the returned pulse. We also have the time taken for the signal to travel to the object and then return, which, when multiplied into the speed of light and divided by two, gives the distance of the object.

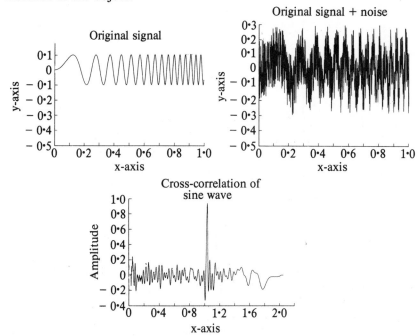

Fig. 5.8   Detection of a swept-frequency signal

Digital Methods for Signal Analysis

It is interesting to consider this process as acting on a spectrally decomposed version of the received signal. This allows us to consider the signal as consisting of a series of frequencies all present at the same time and subject to a range of correlation lags. The output correlation will consist of the summation of all the frequency terms and all the lags, i.e.

$$R_{xy}(\tau,f) = \frac{1}{NB}\int x_f(t)y_f(t+\tau)dt \qquad (5.42)$$

Thus, phase coherence will be obtained for the set of frequencies at certain specific values spaced through the total lag range. This is illustrated in Fig. 5.9 which shows the condition for multiple signal coherence of seven signals subject to seven specific values of lag. It results in a cross-correlation pulse which is short compared with the summation of the seven individual signal periods. At all other lag values expressed by equation (5.42) cancellation of the resultant values will occur.

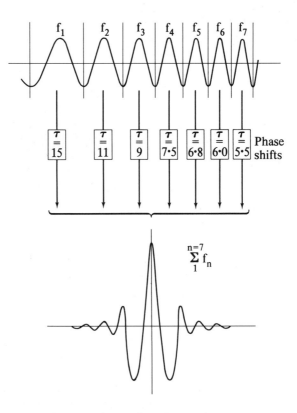

Fig. 5.9   Illustrating phase coherence

Signal extraction by cross-correlation is dependent on the fact that we already know what x(t) is and need only detect its **presence** and amount. Where a signal is not known, but is assumed to be periodic, then the auto-correlation of y(t) = x(t)+n(t) can be used as a method for detecting the period of x(t), since as we saw in Fig. 5.3 R_{yy} will reproduce the periodicity of x(t), whilst the noise, being random, would have no effect on the value obtained. We can then choose a sinusoidal function having the same period as x(t), and use it as a first approximation to x(t). This is then cross-correlated with y(t). The resulting cross-correlation, as we saw earlier, gives a much higher signal-to-noise ratio than R_{yy}, and thus provides us a better estimate about the exact periodicity present in y(t), and also indicates its **phase** (see equation 5.26). The sinusoidal function, with the detected amplitude and phase, is then subtracted from y(t), and the residual signal subjected to a repeated analysis to detect the remaining periodicity. We can then successively determine the most important, the second most important, etc., periodic components in y(t). The total of these components is the desired signal x(t), whilst the remainder constitutes random noise.

We saw earlier that the signal-to-noise ratio of the cross-correlation is $(1+2r+r^2)/(1+r)$ times better than that of the auto-correlation. For large r the ratio is approximately equal to r. Now, after we have determined the most important periodic component of x(t) and subtracted this from y(t), the remainder will have less signal, and hence a worse signal-to-noise ratio, or larger r. However, the improvement of the ratio given by cross-correlation is larger too, and the method may continue to work for several extractions. It is also useful to note that signal-to-noise ratio in R_{yy} and R_{xy} improves with T, hence the longer are the functions available for analysis, the more successful we can expect to be. If, after we perform the analysis, we find that the computed results are too erratic to be of use, indicating that their signal-to-noise ratio is not good enough, it will be necessary to increase T, i.e., obtain more data.

5.8.2 SIGNAL AVERAGING*

An enhancement of the signal-to-noise ratio of repetitive signals can be achieved by making use of the redundant information inherent in the repetition. This is by no means a new technique. (Long exposure photography is an early example used in recording the light signals emanating from distant stars.) This technique may be applied to signal time-histories by summation of successive repetitions of the signal. Although this is not a correlation method, its intent is similar to the method we have just discussed, and is considered here.

The signal may be available in real time form as a continuous periodic signal immersed in noise, or more generally, as a single finite ensemble of periodic or transient form, repeatedly generated from a storage device such as a magnetic tape loop or digital core storage. Successive repetitions of the wave-form are summed so that the periodic signals add coherently, whilst the random element is averaged to a small value by virtue of its incoherence. The technique is known as 'time averaging', and yields an improvement in the signal-to-noise ratio roughly proportional to the square root of the number of repetitions of the signal. This is a useful practical rule having few exceptions and can be assumed where the wave-form of the signal is independent of the repetition rate. Ernst [8] has considered the deviation from this general assumption which can occur under certain conditions.

It is important to note, in connection with signal averaging, that the method is valueless in removing unidirectional trends or very slow changes in the signal base line. To some extent this is applicable to the case of the very low noise frequencies where the noise is inversely proportional to frequency (l/f noise), although some improvement in signal-to-noise ratio for these very low frequencies can be obtained using the multi-sweep technique described below.

The derivation of the signal-to-noise ratio, r_s, of a sampled wave-form will now be considered. The noisy signal, $y(t)$, can be represented as before by $y(t) = x(t) + n(t)$. Two assumptions will now be made which are valid for most noise situations:

1. That the mean value of the noise is zero.
2. That the variance of the noise, σ_n^2, is constant over the duration of the signal, T.

Under these conditions the mean power $\overline{n^2}$ will be equal to the variance, and the signal-to-noise ratio is

$$r_s = (\overline{x^2}) / \sigma_n \tag{5.43}$$

If we sum M repetitions of the signal we have

$$\sum_{k=1}^{M} y_k(t) = Mx(t) + \sum_{k=1}^{M} n_k(t) \tag{5.44}$$

The signal power is now M^2 times larger, but as noise is random the mean square of the sum of M noise terms is only $M\sigma_n^2$. Thus, the signal-to-noise ratio after M repetitions will be

$$r_s(M) = M\sqrt{(\overline{x^2})} / (\sqrt{M})\sigma_n = (\sqrt{M})r \tag{5.45}$$

There is thus an improvement by a factor of $\sqrt{M}$ in the signal-to-noise ratio.

There are two basic ways of carrying out this signal averaging of continuous repeating wave-forms in the time domain. The first is known as **multi-sweep averaging** and is identical to the mathematics described above, i.e., a number of samples of the signal are taken uniformly along its time history and stored in separate locations during the first repetitions of the wave-form. Subsequent sweeps across the signal select the same number and relative time location of the samples and add the value of these to the first set (Fig. 5.10). The second method is known as **single-sweep averaging**.

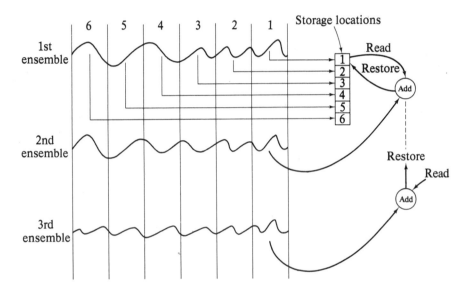

Fig. 5.10   Multi-sweep averaging

Samples are taken after the same delay relative to the commencement of the signal with each repetition of the wave-form. These samples are summed and averaged to produce one final averaged point relative to a particular delay. The delay is then increased by one increment to give a new reference point and the process repeated (Fig. 5.11).

With multi-sweep averaging all the samples of the wave-form are averaged simultaneously so that an average of the complete wave-form is available continuously. Thus, the signal-to-noise ratio is improved with each sweep of the wave-form. This type of averaging is effective in enhancing the signal-to-noise ratio for a large class of noisy signals, including low frequency (l/f) noise. This is because there is a wide separation between n(t), n(t+T), n(t+2T), etc., so that they are likely to be independent, even though n(t) may change fairly slowly and n(t) may be highly correlated with n(t+Δ).

Single-sweep averaging is ineffective with low frequency noise since the noise component at each successive sample of the transient wave-form will

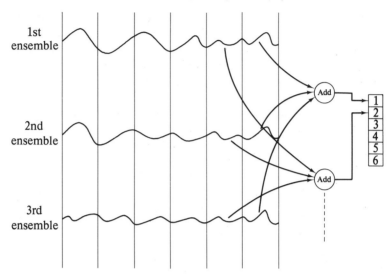

Fig. 5.11   Single-sweep averaging

be highly correlated. Consequently the mean level of the sampled wave-form will vary as the sampled delay is increased and during the slow sweep of the delay reference point this change will be reflected in the final averaged wave-form. It is, however, fairly easy to implement and permits a long integration time per sample (since the summation for a given sample is carried out in parallel). In contrast, the multi-sweep method requires counting of the sweeps and averaging by a different number for each sweep. This demands time-consuming logic calculations between each sample so that the process is generally fairly slow in implementation unless complex fast logic hardware is included.

The time resolution for multi-sweep averaging is determined by the total number of samples taken from the wave-form and is consequently limited to the analysis of fairly slow transients. Single-sweep averaging can have time resolution considerably shorter than this if long sweep times are acceptable. The maximum frequency determined by the multi-sweep aver-ager is limited by the sampling theorem, in contrast to that of the single-sweep averager, where the limit is determined by the width of the sampling pulse.

5.9   Detection of Echoes

The auto-correlation method is, as we have seen, an excellent detector for the periodic components of a composite signal containing a large random noise element. If the signal is of this form and the periodic components are few in number then this technique will give clear unambiguous results. However, much practical data, such as vocal signals and those deriving from periodic transient signals, will have the repeating form shown in Fig. 5.12 and be exceptionally rich in harmonics of the repetition frequency. Consequently although the auto-correlation will indicate the periodic components it will not easily discriminate between their individual contributions. What happens is that the higher frequency resonance will pass in and out of coherence many times as the time lag τ increases, while the lower frequency harmonic components will undergo a smaller number of coherence changes. The result is a composite wave-form showing the modulation of harmonic components by each other which tends to obscure their periodicities. A Fourier transform applied to the result to determine the periodicities will bring us back to the power spectrum indicating relative frequency rather than the rate of frequency change needed to separate the harmonics.

Let us consider first the following simple problem: suppose the signal $x(t)$ is mixed with its own echo, $x(t + s)$, so that the composite signal is $y(t) = x(t) + x(t + s)$. Fourier transforming y produces

$$Y(f) = \int \exp(j\omega t)[x(t) + x(t + s)]dt$$

$$= X(f) + \{\int \exp[j\omega(t + s)]x(t + s)dt\}\exp(-j\omega s)$$

$$= X(f) + X(f)\exp(-j\omega s) \tag{5.46}$$

where $\omega = 2\pi f$.

Thus, the power spectrum of y is

$$S_y = |Y(f)|^2 = |X(f)|^2[1 + \exp(j\omega s)][1 + \exp(-j\omega s)]$$

$$= S_x[2 + 2\cos(\omega s)]$$

Now suppose we take the logarithm of S_y

$$\log_e S_y(f) = \log_e S_x(f) + \log_e[2 + 2\cos(\omega s)]$$

We see that this is the logarithm of S_x plus a periodic function of f, with periodicity increasing with s, the echo delay. If we now compute the Fourier transform of $\log_e S_y(f)$, the periodicity would show up as a peak at 'frequency' s. Thus, by performing Fourier transformation twice, once on $y(t)$ to obtain S_y, and then on $\log_e S_y$, we obtain a result which permits us to identify the echo delay s. This is called the **cepstrum** method [9].

The above is in fact only a special example of a general class of problems called **deconvolution**, in which we wish to recover $x(t)$ from a filtered version $y(t)$, with

$$y(t) = \int h(t-u)x(u)du \tag{5.47}$$

To see the connection, let us put $h(t-u) = \delta(t-u) + \delta(t+s-u)$.

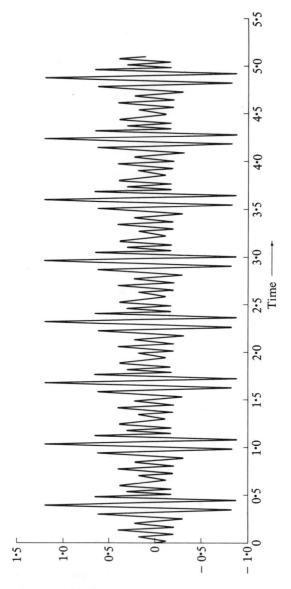

Fig. 5.12   Auto-correlogram of a harmonically rich signal

If we integrate this $h(t-u)$ with $x(u)$ as given in equation (5.47), we would produce $y(t) = x(t)+x(t+s)$. In other words, the echo process may be considered as a special type of convolution, and the recovery of the original signal, $x(t)$ from $y(t)$ amounts to the reverse of the convolution, hence the term, 'deconvolution'.

To study the more general problem, we again Fourier transform y(t). We know that the Fourier transform of a convolution is the product of the transforms involved, or

$$Y(f) = H(f)X(f)$$

and hence, $S_y(f) = |H(f)X(f)|^2 = |H(f)|^2 S_x(f)$

Taking logarithms again gives

$$\log_e S_y(f) = 2\log_e |H(f)| + \log_e S_x(f) \qquad (5.48)$$

In general we know something about the properties of H, e.g. in the echo detection case we know $|H(f)|^2 = 2 + 2\cos(2\pi fs)$ with s being unknown. More commonly, $\log_e |H(f)|$ varies slowly with f whereas $\log_e S_x(f)$ fluctuates more rapidly, and equation (5.48) is similar to the familiar case of slow signal and fast noise mixture, or low frequency/high frequency mixture. As shown in the next section, we can recover the low frequency signal by performing a low-pass filtering, and so, recover $\log_e |H(f)|$. As h(t) is the Fourier transform of H(f), we obtain an approximation to it, and thus gain an understanding of the convolution process. Finally, by filtering y(t) using a filter whose impulse response function is $1/H(f)$, we can recover x(t) from it. Here we have the basic idea of deconvolution. However, the technical details are rather complex, and one must do a great deal of experimentation to achieve satisfactory results. We shall not discuss the topic further here. One example, in picture de-blurring, is given a brief mention in Chapter 7. Another is the problem of making accurate estimates of the firing rate from an examination of the auto-correlogram of an acoustic signal derived from an automobile engine [10]. This is shown in Fig. 5.12 reproduced from Dr Thomas's paper. The results of applying cepstrum analysis to the same repeated transient signal are shown in Fig. 5.13 and indicate quite clearly the signal corresponding to the engine firing rate at a position well removed from random noise background.

The technique described above is called 'blind' deconvolution because we only have y(t) to work with. x(t) has not been available. In the next section we shall see a different sort of problem, where x(t) and y(t) are both available for analysis.

5.10   System Testing

Mechanical systems, particularly complex cyclic mechanisms, are often studied from the characteristics of the vibrations set up by the system during normal operations. Two general approaches to this study are common. The first involves signature analysis in which the characteristics of a vibration time-history are compared during the operational life of the

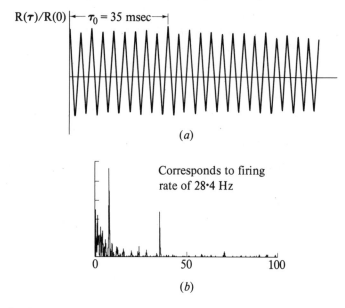

(a)

(b)

Fig. 5.13   Comparison between (a) the auto-correlogram and (b) cepstrum
analysis obtained from the same repeated transient

mechanism to determine departure from 'normal' signature, obtained using
a process of time averaging [11]. The second uses cross-correlation techniques to determine the transfer function, related to the mechanical impe-
dance, of the mechanism [12]. Time averaging has been discussed in the
previous section and, as applied to a group of signals obtained from a
mechanism which is supposed to be functioning normally, will enable small
changes in performance to be detected before failures occur [13]. The
technique can therefore be used to predict and diagnose malfunction.

Comparison between the normal signature and that obtained during later
operation can use statistical techniques, spectral density analysis of the
difference wave-forms, or the process of cross-correlation. This latter
technique is also used in the determination of system transfer function. A
major advantage of correlation techniques is that they can be applied
without disturbing the normal operation of the system. Earlier methods
using classical Bode plotting of the amplitude and phase characteristics
from a variable frequency drive generator have a number of disadvantages,
not least of which is the need to carry out these measurements off-line. This
is also true of impulse-response methods which, although considerably
faster than the Bode plot, introduce problems of implementation. The
method of impulse-response testing is to determine the behaviour of the
system to a single Dirac delta function applied as its input. The delayed
impulse response produced at the output can be analysed to give the

frequency amplitude and the phase response of the system. Unfortunately it is rarely possible to investigate the characteristics of the system in this way. Either the pulse will result in overload conditions, which modifies the behaviour from the operational behaviour, or, to avoid this, the impulse is reduced to such a small amplitude that the output produced will be masked by the system noise. It is also extremely difficult to maintain the energy spectrum of the impulse constant over the system bandwidth.

We saw earlier that the auto-correlation of a wide-band noise signal is approximately an impulse function, and this fact can be used in an equivalent form of impulse testing. The method is shown in Fig. 5.14. If a

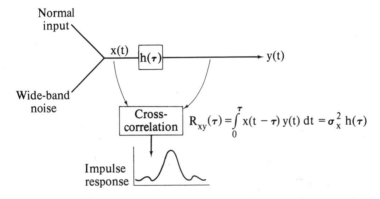

Fig. 5.14 Determination of system impulse response by cross-correlation

wide-band noise signal is applied to the system input, **the cross-correlation of this with the output will be proportional to the impulse-response function of the system.** Mathematically, if $x(t)$ is the input noise signal and $h(v)$ is the required impulse-response function, then the output signal is

$$y(t) = \int h(v)x(t-v)dv \qquad 5.49)$$

Cross-correlating $y(t)$ with the input $x(t)$ gives

$$R_{xy}(\tau) = <x(t)y(t+\tau)> = \int h(v)<x(t)x(t+\tau-v)>dv$$

$$= \int h(v)R_{xx}(\tau-v)dv \qquad (5.50)$$

This is known as the Wiener–Lee relation. It states that the cross-correlation between the input and the output is the convolution of the impulse response with the auto-correlation of the input. Since the auto-correlation of a wide-band noise process is just a delta function, we can write

$$R_{xy}(\tau) = \Lambda \int h(v)\delta(\tau-v)dv = \Lambda h(\tau) \qquad (5.51)$$

Since $\delta(\tau-v) = 0$ everywhere except at $\tau = v$, the only value to remain is that of $h(\tau)$. Thus, we see that measuring the cross-correlation R_{xy} immediately produces the impulse response for the system.

Even when we are unable to control the input, equation (5.50) still gives a useful method for determining the impulse response h. If we take the Fourier transform for both sides of equation (5.50) we have

$$\int R_{xy}(\tau)\exp(-j\omega\tau)d\tau = \int \exp(-j\omega\tau) \int h(v)R_{xx}(\tau-v)dvd\tau$$

The left-hand side is, by definition, $S_{xy}(f)$. The right-hand side can be written as

$$\int \exp(-j\omega v)h(v) \int \exp[-j\omega(\tau-v)]R_{xx}(\tau-v)d\tau dv$$

$$= H(f)S_x(f) \tag{5.52}$$

Thus, in the frequency domain we simply have $S_{xy}(f) = H(f)S_x(f)$. Consequently, given any general input function x(t) and its corresponding output y(t) we can extract information about h(t) by computing $H(f) = S_{xy}(f)/Sx(f)$, from which we obtain h simply by an inverse transformation. This is how we can test the system during its normal running.

Now let us consider a somewhat more general system than that given by equation (5.49). Suppose

$$y(t) = \int x(t-v)h(v)dv + n(t) \tag{5.53}$$

where n(t) is the random noise generated by the system.

Under certain conditions, where x(t) and n(t) are independent random processes, equation (5.50) remains valid. To demonstrate this, we take

$$R_{xy}(\tau) = \int h(v)<x(t)x(t+\tau-v)>dv + <x(t)n(t+\tau)>$$

Since x(t) and n(t) are assumed to be independent, the second term is identically zero, and we simply have equation (5.50). Thus, we can still apply the formula $H(f) = S_{xy}(f)/S_x(f)$. Further, we can also derive some information about the noise process. If we correlated each side of equation (5.53) with $y(t+\tau)$ we get

$$R_y(\tau) = <y(t)y(t+\tau)> = <y(t)[\int h(v)x(t+\tau-v)dv+n(t+\tau)]>$$

$$= \int h(v)R_{xy}(v-\tau)dv + <n(t)n(t+\tau)>$$

$$= \int h(v)R_{xy}(v-\tau)dv + R_n(\tau)$$

Taking the Fourier transform on both sides we would get

$$S_y(f) = H(f)S_{xy}(f)^* + S_n(f) = S_{xy}(f)^*S_{xy}(f)/S_x(f) + S_n(f)$$

or
$$S_n(f) = S_y(f)[1 - |S_{xy}(f)|^2/S_x(f)S_y(f)] \tag{5.54}$$

By definition, $|S_{xy}(f)|^2/S_x(f)S_y(f)$ is $\kappa_{xy}(f)$, the coherency spectrum. Thus

$$S_n(f) = S_y(f)[1-\kappa_{xy}(f)] \tag{5.55}$$

We can thus derive the noise spectrum from S_x, S_y and S_{xy}.

It is instructive to note that, where $\kappa_{xy} = 1$ $S_n = 0$, while where $\kappa_{xy} = 0$ $S_n = S_y$. In other words, the coherency spectrum between input x(t) and output y(t) gives us a measure of the signal-to-noise composition of y(t). Those frequencies at which $\kappa \simeq 1$ occur where output is due mainly to the input, while those at which $\kappa \simeq 0$ occur where the system-generated noise predominates. Thus, correlation/spectral methods, applied appropriately, are powerful methods for understanding the behaviour of linear systems.

5.10.1 PSEUDO-RANDOM NOISE

The characteristic of the random noise required for the system testing method, where we use a wide-band noise as a test input x(t) for a system, is that its auto-correlation be a delta function, $\delta(\tau)$. This property implies that the noise contains equal amounts of all frequencies, and hence gives rise to the name 'white noise' in analogy to white light. True white noise cannot be realised since an infinite bandwidth would be needed. It is generally sufficient to specify a level power spectrum over the frequency band of interest. Generation of a random noise signal can be obtained by utilising the random behaviour of gaseous discharges of semi-conductor materials. These natural methods of noise generation have one major disadvantage, namely their unrepeatable behaviour. It would be necessary to average the result over a very long interval to establish a small variance value. This will be the case for low frequency measurements and results in long computational times.

A test signal which avoids these difficulties consists of a repeated pattern of random noise having a predetermined power spectrum and amplitude probability distribution. This is known as pseudo-random noise. If the measurement time is made equal to a multiple of the pattern length then a variance of zero will be achieved resulting in short measurement times and high statistical accuracy. Pseudo-random sequences are generated as a two level signal using shift register techniques [14, 15]. If an analog pseudo-random signal is required then low-pass filtering of the binary sequence is carried out. There also exist methods for generating random numbers by software and converting these to an electric signal by digital-to-analog conversion [16].

The auto-correlation function of a pseudo-random binary noise signal is triangular with a base-width equal to two sampling periods (Fig. 5.15). It thus forms a close approximation to the Dirac delta function required, and is thus a reasonable input noise generator for the testing method discussed earlier. The fact that the signal is actually repeatable, or periodic, is not apparent in the correlated signal, providing that the period is sufficiently long.

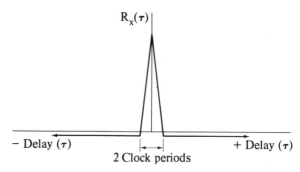

Fig. 5.15   Auto-correlation of pseudo-random noise

5.11   Auto-Correlation by Walsh Transform*

Earlier we discussed digital methods for estimating the auto-correlation function of a random process, directly or via the fast Fourier transform. Two relatively recent papers [17] [18] proposed a new, faster method for this purpose. The method is based on the relation between the auto-correlation and the Walsh power spectrum of a random process. As we said in Section 4.7.3, the Walsh power spectrum is easily estimated and requires only $N \log_2 N$ real additions to compute. A fast Walsh transform on the spectrum, requiring another $N \log_2 N$ additions, produces what one calls the **logical** auto-correlation function, which is related to the normal auto-correlation function by a simple matrix multiplication. It is thus suggested that the three-step method is an efficient technique for auto-correlation estimation.

Unfortunately, our study shows that the method produces very poor results. While the auto-correlation estimate produced this way is **on average** equal to the theoretical auto-correlation, or, it is an **unbiased** estimate, this alone does not make it a good estimate, which requires that the **variance** be small. It is on this latter requirement that the method falls down. We can show this theoretically for certain random processes, but will not devote any space to the analysis here. Instead, we shall demonstrate this by an example and provide some brief explanation on the cause of the problem.

The results presented here were produced as follows: We first generated 65 vectors of independent Gaussian random numbers, with each vector containing 256 elements. These were filtered recursively using the formula

$$x_i = n_i + x_{i-1} - {}^1\!/_2 x_{i-2}$$

(also used in Section 4.6.5), n being an element of the random vector being processed. The input random process has a known auto-correlation, shown in Fig. 2.10. For each of the vectors we estimated the first 32 values of the auto-correlation by the direct method and then by the Walsh spectrum method. The variance of each computed result was then computed by averaging over the 65 vectors.

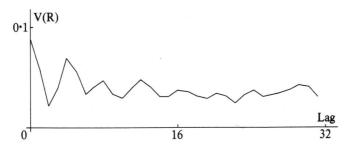

Fig. 5.16   Variance of auto-correlation estimate by direct method

Figure 5.16 shows the 32 variances of the direct method results. It is seen that, except for R(0), with a variance double of that of the rest, the variance plot is approximately constant. Figure 5.17 shows 32 variances from the Walsh transform method. The plot fluctuates wildly. A few good elements have variances quite comparable to those from the direct method, but generally the variances are much larger, some by several orders of magnitude.

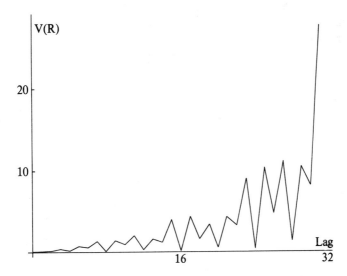

Fig. 5.17   Variance of auto-correlation estimate by Walsh transform method

The problem has its cause in the last step of the Walsh transform method. The conversion of the logical auto-correlation to the normal auto-correlation is by the multiplication of a matrix described by mathematicians as being 'ill-conditioned'. Expressed in simple terms, any arithmetic process that adds numbers of similar values and divides the total by a large number, as in taking averages, is well conditioned and reduces variance. Any arithmetic process that subtracts similar numbers and then multiplies the result by a large number is ill-conditioned and magnifies variance. Because the Walsh transform method used such a matrix to convert the logical auto-correlation to the normal auto-correlation, the final result has large variance, and is not a good estimate. Consequently, despite its speed advantage the method must be rejected.

References

1. WAINSTEIN, L. A., and ZUBAKOV, V. D. *Extraction of Signals from Noise*. Prentice-Hall, 1962.
2. ROBINSON, E. A. *Multi-channel Time Series Analysis with Digital Computer Programs*. Holden-Day, San Francisco, 1967.
3. SCHMID, L. P. Efficient auto-correlation. *Comm. ACM,* **8**, 115 (1965).
4. KORN, G. A. *Random Process Simulation and Measurements*. McGraw-Hill, New York, 1966.
5. VAN VLECK, J. H. The spectrum of clipped noise. *Rep. 51, Radio Research Laboratory,* Harvard University, 1943.
6. WEINREB, S. A. A digital spectral analysis technique and its applications to radio astronomy. *M.I.T. Research Lab. Electronics Tech. Rep.,* 412, 1963.
7. LEE, Y. W., CHEATHAM, T. P., and WIESNER, J. B. Application of correlation analysis to the detection of periodic signals in noise. *Proc. IRE,* **38**, 1165–71, 1950.
8. ERNST, R. R. Sensitivity enhancement in magnetic resonance I. Analysis of the method of time averaging. *Rev. Sci. Instru.,* **36**, 12, 1689–95, 1965.
9. BOGERT, B. P., HEALEY, M. J. R., and TUKEY, J. W. The quefrequency analysis of time series for echoes, cepstrum, pseudo-autocovariance, cross-cepstrum and saphe cracking. In *Proceedings of the Symposium on Time Series Analysis*. Ed. M. Rosenblatt, Wiley, New York, 1963.
10. THOMAS, D. W., and WILKINS, B. R. Determination of engine firing rate from the acoustic wave-form, *Electronics Letters,* **6**, 7, 193–6, 1970.
11. HOFFMAN, R. L., and FUKUNGA, K. Pattern recognition signal processing for mechanical diagnostics signature analysis. *IEEE Trans. Comput.,* **C-20**, 1095–100, 1971.
12. MERCER, C., and CLARKSON, B. C. Use of cross-correlation in studying the response of lightly damped structures to random forces. *J. AIAA,* Dec. 1965.
13. LAVOID, F. J. Signature analysis: Product early warning system. *Mach. Des.,* **41**, 151–60, 1969.
14. GILSON, R. P. Some results of amplitude distribution experiments on shift register generated pseudo-random noise, *IEEE Trans. Comput.,* **C-15**, 926, 1966.

15. HUTCHINSON, D. W. A new uniform pseudo-random number generator. *Comm. ACM,* **9**, 432–3, 1966.
16. BRENT, R. P. Algorithm A488, *Comm. ACM,* **17**, 704–6, 1974.
17. AHMED, N., and NATARAJAN, T. On logical and arithmetic auto-correlation functions. *IEEE Trans. Electromag. Compat.,* **16**, 177–83, 1974.
18. LOPRESTI, P. V., and SURI, H. L. A fast algorithm for the estimation of auto-correlation functions. *IEEE Trans. Audio Speech Signal Proc.,* **22**, 449–53, 1974.

Chapter 6

Digital Filtering

6.1 Introduction

Linear filters are of crucial importance in the processing of signals. Whilst in its most general sense a filter can be considered as a device used for selecting any particular frequency or set of frequencies the term is usually confined to a system which transmits a certain range of frequencies, rejecting all others. Such frequency ranges are called **pass-bands** and **stop-bands** respectively.

A **digital filter** acts on a sampled version of the signal to be filtered and may be realised through discrete logical hardware elements or by suitable programming of a digital computer which is fed with a sampled version of the input signal. This chapter will be concerned primarily with the computer realisation of the digital filter.

The process of digital filtering represents an operation on a discrete series of input values such that the output series produced is dependent on the input series and the modifying coefficients defining the filter characteristics. It is necessary to assume that the process is a linear one so that the principle of superposition applies to the input and output series.

The principal use of the digital filter or numerical filter is to smooth a data series in the time or frequency domain. It will show the characteristics of its analog counterpart, i.e. low-pass, band-pass, high-pass, etc., but can also have some properties not possible for the analog form. As an example we can consider the filter to operate in other than real time so that its response to a unit impulse may not be zero for $t < 0$. It may also be arranged to have zero phase characteristics and accept data in a reverse order.

We can approach the specification of a digital filter via either the **time domain** or the **frequency domain**.

The **impulse-response function** in its discrete form as a number sequence can be used to specify the filter in the **time domain**. The sequence may be used directly to form a convolution of the signal with the response function, or indirectly via a transformation in the frequency domain. The advent of the fast Fourier transform algorithm has played an important role in the indirect filtering operation from the time series which often proves more economical than the direct method.

Frequency specification in terms of amplitude and phase characteristics express the 'classical' method of considering continuous analog filters, and may also be used with discrete digital filters. Since a wealth of design information on the transfer function characteristics of continuous filters is available, synthesis methods have been developed to permit conversion of these characteristics into discrete forms so that the design of a digital filter can proceed.

A definition of a transfer function in terms of its polynomials is widely used for continuous filter representation. Modern network synthesis is based on quite a different representation, that of poles and zeros. Before we consider a response function approach of digital filter design we must first understand the relationship between these two forms of filter representation.

6.1.1 POLE AND ZERO LOCATIONS OF TRANSFER FUNCTIONS

A linear system, such as a filter, can be characterised by a complex transfer function expressed in Laplace form as

$$H(s) = \frac{Y(s)}{X(s)} = \frac{\alpha_0 s^n + \alpha_1 s^{n-1} + \alpha_2 s^{n-2}, \ldots, \alpha_n}{\beta_0 s^m + \beta_1 s^{m-1} + \beta_2 s^{m-2}, \ldots, \beta_m} \tag{6.1}$$

This can be factored into roots of the numerator and denominator polynomials

$$H(s) = K \frac{(s-a_1)(s-a_2), \ldots, (s-a_n)}{(s-b_1)(s-b_2), \ldots, (s-b_m)} \tag{6.2}$$

where a_1, a_2, etc. designate the roots of $Y(s)$ and b_1, b_2, etc. designate the roots of $X(s)$. When the complex frequency $s = j\omega$ assumes any values, a_1, $a_2, \ldots, a_n$, the system response is zero. When it assumes any value, b_1, $b_2, \ldots, b_m$, the response is infinite. These values of s are the **zeros** and **poles** of the transfer function, $H(s)$, respectively.

The substitution of $s = j\omega$ in equation (6.2) will enable the frequency behaviour of the filter to be determined. It enables filter performance to be completely specified by means of a complex frequency diagram (s-plane), in terms of real and imaginary values and a constant filter gain, K. Figure 6.1 shows a typical pole–zero configuration for a continuous stable filter. Poles are represented by crosses and zeros by circles. It can be shown that the criterion for stability is that all the poles must lie in the **left-hand** half of the s-plane. Also since the filter impulse response, as representative of a physical system, is a real function of time, then any unreal poles or zeros must exist in **complex conjugate pairs**. The number of poles must be exactly equal to the number of zeros, although it is possible for certain zeros to be placed at infinity (e.g. if s is very large).

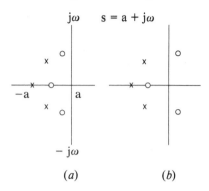

Fig. 6.1   Pole−zero representation of a continuous filter

This representation in terms of poles and zeros can yield the frequency and phase characteristics of the filter either by graphical methods from the s-plane, or by substitution of the complex frequency for s in the factored form of the transfer function. Although the location of poles in the left-hand half of the s-plane is essential for stability, this is not the case for the zeros. The location of zeros is important from the point of view of filter phase response. If in the case of Fig. 6.1(a) the two zeros above and below the line are transferred to mirror-image positions on the right-hand half of the s-plane then the length of the vector zeros drawn to any point on the imaginary axis is clearly unaltered, see Fig. 6.1(b). Their contribution to the phase response is however changed whilst the filter remains stable. Similar changes can be brought about by adding pole−zero pairs in mirror-image positions on either side of the imaginary axis.

If zeros are restricted to the left-hand half of the s-plane, magnitude and phase response are related and the filter is known as a **minimum phase filter**. If, on the other hand, the zeros of the filter transfer function are placed in either half then some flexibility in altering the phase response may be achieved.

6.1.2   DISCRETE FILTER REPRESENTATION

Equation (6.1) is a form of differential equation applicable to the representation of continuous filters. In the discrete case we use the **linear difference equation**

$$y_i = \sum_{k=0}^{P} b_k y_{i-k} + \sum_{k=0}^{M} a_k x_{i-k} \quad (i=1,2,\ldots,N) \tag{6.3}$$

where P and M are positive integers and a_k and b_k are real constants. When M = 0 the filter is auto-regressive and the time duration of the filter impulse response is effectively infinite. This is known as the IIR (infinite impulse response) or **recursive filter**.

For P = 0 the duration of impulse response is limited and the filter is known as the FIR (finite impulse response) or **non-recursive filter**.

The two classes of filter have quite different properties. The FIR filter represents the summation of a limited number of input terms and thus has a **finite memory**. It has excellent phase characteristics but requires a large number of terms to obtain a relatively sharp attenuation characteristic.

The IIR filter represents the summation of both input and output terms so that it can be considered as having an **infinite memory**. It requires relatively few terms for a similar attenuation characteristic, but will possess poorer phase performance.

The appropriate transform for difference equations is the z-transform [1], which performs the same role with difference equations as the Laplace transform carries out for differential equations. The use of the z-transform permits the specification of a digital filter from the continuous transfer function directly in terms of delays, multipliers and adders which is the correct form for hardware filter implementation [2]. Its use for 'software' filters permits a convenient form of expression for the difference equations derived from the general form of equation (6.3), particularly when IIR filters are considered. It is of less value when used in the implementation of software FIR filters since these are inherently stable, and generally imply the use of Fourier transform methods.

In summary we can recognise three fundamental techniques used in digital filtering. These are

1. convolution (direct filtering);
2. Fourier transformation (indirect filtering);
3. auto-regression (use of difference equations).

The first two techniques are generally confined to the FIR filter and the third to the IIR filter.

Essentially the synthesis of a digital filter follows the form familiar with continuous (analog) filters. That is, we need to choose the ideal desired characteristic in the frequency domain, determine an acceptable approximation for this and finally synthesise the filter by the calculation of the filter weights.

Before filter synthesis is considered in detail the behaviour of filters in the time and frequency domain will be described.

6.1.3 THE IMPULSE-RESPONSE FUNCTION

Considering first filter operation in the time domain, we can regard the filter as a 'black box' having accessible input and output terminals. The charac-

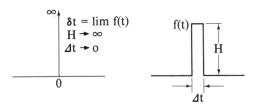

Fig. 6.2   The Dirac impulse function

teristics of such a system are determined precisely by applying to its input terminals a known function and measuring the function appearing at its output terminals. Let this input be a **Dirac impulse function** δt, (Fig. 6.2) defined as

$$\delta(t) = \lim_{\substack{H\to\infty \\ \Delta t\to 0}} f(t) \tag{6.4}$$

where the integral of $f(t)$ is always unity, i.e.

$$\int_{-\infty}^{\infty} f(t)\,dt = \lim_{\substack{H\to\infty \\ \Delta t\to 0}} H\Delta t = 1 \tag{6.5}$$

The total response of the system to this unit impulse is determined by the system characteristics, expressed as a weighting function h(t), which in this case can be termed the **impulse-response function**. h(t) uniquely defines the system and will always follow the same form for a Dirac input function. Thus for a second-order damped system corresponding to a narrow-band resonant filter the impulse response would be the decaying wave-form shown in Fig. 6.3

This argument can be extended to any arbitrary wave-form, x(t), applied to the system by using the principle of superposition. Referring to Fig. 6.4, we assume that the driving wave-form consists of a series of narrow samples, each representing a delta function multiplied by a coefficient proportional to the height of the curve. This is a direct consequence of equation (6.4) since the integral of a function multiplied by δt must be equal to the function itself at t = 0. Thus we can define the output wave-form as

$$x(t) = \int_{0}^{t} x(\tau)\,.\,\delta(t-\tau)\,.\,d\tau \tag{6.6}$$

where x(τ) represents the wave-form coefficient at t = τ and, since the unit impulse response of a physically realisable linear system is zero for negative values of τ, then we can replace the infinite lower limit of equation (6.5)

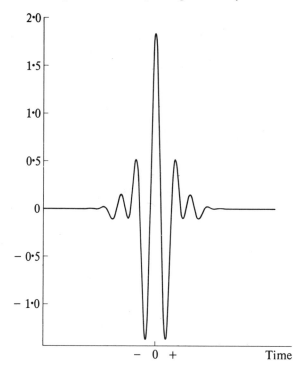

Fig. 6.3   An impulse-response function

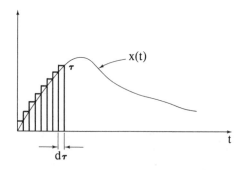

Fig. 6.4   Synthesis of a driving wave-form

by zero. The response function to this arbitrary input wave-form, x(t), will be the total sum of the input response functions for each individual unit impulse

$$y(t) = \int_{-\infty}^{\infty} x(\tau) \cdot h(t-\tau) \cdot d\tau \qquad (6.7)$$

We can consider equation (6.7) to be obtained from the product of the instantaneous value of x(t), namely x(τ) and its weighting function, h(t$-\tau$), together with current values of impulse-response functions initiated earlier. The output function y(t) expresses the convolution product x(t) and h(t) and can be written

$$y(t) = x(t)*h(t) \qquad (6.8)$$

Equation (6.7) represents a convolution integral and may also be written

$$y(t) = \int_{-\infty}^{\infty} h(\tau)x(t-\tau).d\tau \qquad (6.9)$$

The weighting function of time, h(τ), describes the system characteristics in the time domain and, as included in equation (6.9), would consist of a continuous function. A discrete form of equation (6.9) will be considered in Section 6.2 where it is used to implement the FIR digital filter.

6.1.4 THE FREQUENCY-RESPONSE FUNCTION

An alternative, and perhaps more familiar way of describing a linear network system such as a filter, is via a frequency domain representation. If we again consider a black box representation but this time consider the input to be an arbitrary frequency function X(ω), where X(ω) is the Fourier transform of x(t)

$$X(\omega) = \int_{-\infty}^{\infty} x(t)\exp(-j\omega t).dt \qquad (6.10)$$

then the output frequency response will be determined by the system transfer function H(ω), where (ω) will be a complex quantity. Since the transfer function also uniquely defines the system in the frequency domain, we would expect a relationship to be obtained between this and the impulse-response function, h(t), in the time domain. They are, in fact, related by a Fourier transform pair, viz.

$$H(\omega) = \frac{1}{2\pi} \int_{-\infty}^{\infty} h(t)\exp(-j\omega t).dt \qquad (6.11)$$

and

$$h(t) = \int_{-\infty}^{\infty} H(\omega)\exp(j\omega t).d\omega \qquad (6.12)$$

The output of the system, y(t), can be obtained by taking the Fourier transform of equation (6.7) viz.

$$Y(\omega) = \frac{1}{2\pi} \int_{-\infty}^{\infty} \exp(-j\omega t) . dt \int_{-\infty}^{\infty} x(\tau)h(t-\tau) . d\tau$$

If we let the variable of integration be $p = t - \tau$ then

$$Y(\omega) = \frac{1}{2\pi} \int_{-\infty}^{\infty} \exp[-j\omega(p+\tau)]1 . dp \int_{-\infty}^{\infty} x(\tau)h(p) . d\tau$$

$$= \left[\int_{-\infty}^{\infty} x(\tau)\exp(-j\omega\tau) . d\tau \right] \left[\frac{1}{2\pi} \int_{\infty}^{\infty} h(p)\exp(-j\omega p) . dp \right]$$

We recognise the bracketed terms as the Fourier transforms of the input function, $X(\omega)$, and the impulse-response function, $H(\omega)$, respectively which enables the important relationship between them to be stated as

$$Y(\omega) = X(\omega) . H(\omega) \qquad (6.13)$$

This may be compared directly with equation (6.8). We note here the significant result that a product in the frequency domain transforms to a convolution in the time domain.

This relationship provides a method of obtaining filtering action using the Fourier transform of equation (6.12), thus

$$
\begin{array}{ccccc}
y(t) & = & x(t) & * & h(t) \\
\uparrow & & \downarrow & & \downarrow \\
| & & DFT & & DFT \\
| & & \downarrow & & \downarrow \\
IDFT & & X(\omega) & .H(\omega) = & Y(\omega)
\end{array}
\qquad (6.14)
$$

This method is particularly valuable with digital computation, since it proves more economical in computing time to obtain a convolution of two time series by the product of their transforms and subsequent inverse transformation, than the direct product method of evaluating summations.

It should be noted that the transfer function defined by equation (6.11) is a complex quantity, $H(\omega) = a(\omega) + jb(\omega)$, where $a(\omega)$ represents the real part and $b(\omega)$ the imaginary part. Consequently the amplitude characteristic of a smoothing filter defined from $H(\omega)$ is given as

$$| H(\omega)| = [a^2(\omega) + b^2(\omega)]^{1/2} \qquad (6.15)$$

and the phase characteristic by

$$\phi(\omega) = \arctan [b(\omega)/a(\omega)] \qquad (6.16)$$

6.2   The FIR Digital Filter

Ideal filter representations in the frequency domain are shown in Fig. 6.5. The approximation problem in filter design is to decide upon a suitable criterion or set of criteria which will permit a satisfactory approximation to these ideal characteristics. Whatever criterion is selected, several basic limitations to the filter model will be found. With digital filter design, the most common limitation is that, for a given number of filter weights, a steep change in slope is likely to be accompanied with considerable frequency ripple in the realised filter shape corresponding to large amplitude side lobes present in its impulse-response function. The FIR digital filter consists of the summation of the products of M weighting coefficients h_k and the present and past N samples of the signal wave-form

$$y_i = \sum_{k=0}^{M-1} h_k \cdot x_{i-k} \quad (i = 1,2,\ldots,N; k = 0,\ldots,M - 1) \qquad (6.17)$$

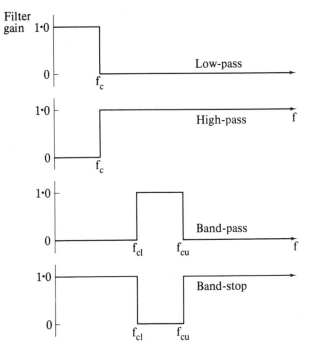

Fig. 6.5   Ideal filter characteristics

which will be recognised as the discrete sample equivalent to the continuous convolution equation of (6.9). The phase characteristics of such a filter are linear and under some circumstances can be made constant at all frequencies. The problems associated with frequency aliasing will be present in filter design and in order to avoid these it is necessary to ensure that the signal sampling rate is at least twice that of the highest frequency of interest, f_s, applied to the filter.

The numerical filter itself consists of a set of fixed weights, h_k, and the process of filtering consists of the determination of the sum of products, $h_k \cdot x_{i-k}$, from the set of filter weights at each sampled data value. This is illustrated in Fig. 6.6 where the set of filter weights, h_k, form an impulse-response function over a time interval smaller than the time period of the signal, x_i. For convenience the function h_k is assumed to move along the time axis from left to right. At each point along the time axis the average product of $h_k \cdot x_i$ is determined so that the entire process can be considered as the production of a 'moving average' of the data series as the filter is moved along it. Note here that the discrete impulse-response function, defined by h_k, does not have to be zero for $\tau < 0$ as with a real physical system. We shall see later that the determination of filter weights via the Fourier transform commences with a two-sided response in the frequency domain and results in an impulse-response function, or series of filter weights, symmetrically disposed about $\tau = 0$.

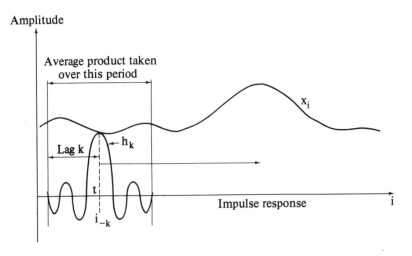

Fig. 6.6   Evaluation of a moving time average

The moving average value produced by the process of convolution described above will be dependent on the frequency content of the data time-history. As an example we may consider the representation of a low-pass filter where the filtered output will be unchanged (i.e. the con-volved value, $\Sigma h_k \cdot x_{i-k}$ will be unity) for frequencies below cut-off fre-quency. Above this frequency the time-history will contain several alterna-tions and the incremental convolved values, $h_k * x_{i-k}$, will have alternative signs and will therefore tend to cancel during the summation process. The frequency characteristic of the filter may be derived from equation (6.12), replacing the filter transfer function by its discrete Fourier transform

$$Y(\omega) = X(\omega) \sum_{k=-M}^{M} h_k \exp(-j\omega kT) \tag{6.18}$$

where T is the constant time interval between successive samples. We can write for the transfer function of the filter

$$H(\omega) = \frac{Y(\omega)}{X(\omega)} = \sum_{k=-M}^{M} h_k \exp(-j\omega kT) \tag{6.19}$$

This is a complex number and can be expressed as

$$H(\omega) = h_k \cdot \cos\theta(k) + jh_k \cdot \sin\theta(k) \tag{6.20}$$

where the filter gain is $|h_k|$ and the phase shift is $\theta = \omega T$. If we now let $h_k = h_{-k}$ for all the values of k from $k = 0$ to $k = M$ (equivalent to a phase shift of zero or π radians) then the transfer function will become a real number for any value of k and we can write

$$H(\omega) = h_0 + 2. \sum_{k=1}^{M} h_k \cdot \cos \omega kT \tag{6.21}$$

which defines a cosine FIR filter having zero phase shift. Similarly, for $h_{-k} = -h_k$ (equivalent to a phase shift of $+ n/2$ radians), then $H(\omega)$ will become purely imaginary and will define the sine FIR filter

$$\frac{H(\omega)}{j} = 2. \sum_{k=1}^{M} h_k \cdot \sin \omega kT \tag{6.22}$$

This possibility of obtaining zero or fixed phase shift at all frequencies is a particularly valuable feature of digital filters. The discrete Fourier transform used in the derivation of equations (6.21) and (6.22) represents a finite approximation to the infinite integral of equation (6.11). As a conse-quence the Fourier series will not converge smoothly and errors will result from the truncation of the infinite series. Methods of minimising this error by modification to the Fourier series will be considered in Section 6.2.2 when the derivation of weights for the FIR filter is discussed.

6.2.1 USE OF THE FAST FOURIER TRANSFORM

The direct method of evaluating the integral given by equation (6.17) proves uneconomic when large numbers of sample values are considered so that the indirect method is preferred. This makes use of the equivalence of a product in the frequency domain and a convolution in the time domain, given by equation (6.14). Here the fast Fourier transform can be used to reduce substantially the amount of computation time required.

To carry out convolution in this way the signal and the impulse-response function are transformed into the frequency domain, their products formed and the new series inversely transformed back into the time domain to provide a filtered version of the original signal. Unfortunately, due to the need for handling complex numbers at all stages of these operations and the need to avoid circular convolution, the length of data that can be handled in the main storage of a digital computer is limited. This is due partly to the need to store a complex variable in two- or four-digital words and to the Cooley–Tukey FFT algorithm which requires that the length of the data be a power of two. Additionally to avoid circular convolution, zeros, equal in number to the length of the signal, need to be inserted in the signal series.

Fortunately certain techniques are available which permit the convolution of a long data series by a process of segmentation and continuous convolution which uses less in-core storage. These are known as the **select-save** [3] and **overlap-add** [4] methods. The first of these is described here.

If we consider the signal series, x_i, subject to a filter series of length, h_k, where $i = 0,1,\ldots,(N-1); k = 0,1,\ldots,(M-1)$ and $M \ll N$ then equation (6.17) (repeated below)

$$y_i = \sum_{k=0}^{M-1} h_k \cdot x_{i-k}$$

represents a difference equation describing the FIR filter. A procedure will now be evolved which considers the repeated convolution of a modified filter series, h_ι, with the total signal series, x_i, divided into a number of smaller series, x_m, each of identical length to the modified filter series, h_ι. The addition of these fractional convolutions will be shown to be equivalent to the complete convolution given by equation (6.17). If we consider a signal series, x_m, where $m = 0,1,\ldots,(L-1)$, and $L < N$, and a modified filter series, h_ι, where $\iota = 0,1,\ldots,(L-1)$ then if

$$X_n = \text{DFT of } x_m = \frac{1}{L} \sum_{m=0}^{L-1} x_m \exp(-j2\pi mn/L) \qquad (6.23)$$

and

$$H_n = \text{DFT of } h_\iota = \frac{1}{L} \sum_{\iota=0}^{L-1} h_\iota \exp(-j2\pi\iota n/L) \qquad (6.24)$$

we can write the convoluted product of x_m and x_ι as the IDFT of $X_n \cdot H_n$, i.e.

$$y_p = \sum_{n=0}^{L-1} X_n H_n \exp(j2\pi pn/L) \qquad (6.25)$$

Substituting for X_n and H_n and rearranging the summations

$$y_p = \frac{1}{L^2} \sum_{m=0}^{L-1} \sum_{\iota=0}^{L-1} x_m h_\iota \sum_{n=0}^{L-1} \exp[k2\pi n/L(p-\iota-m)] \qquad (6.26)$$

but

$$\sum_{n=0}^{L-1} \exp[j2\pi n/L(p-\iota-m)] = L \text{ if } (p-\iota-m) = qL = 0 \text{ otherwise}$$

where q is an integer. This follows from the principle of orthogonality. Now since p,l,m vary in the range 0 to $L-1$ then q can only be 0 and -1, so that $p-\iota-m = 0$ and $\iota = p-m$, $p-\iota-m = -L$ and $\iota = L+p-m$ are the two possible ranges for the variable h_ι, and we can substitute these in equation (6.26) and sum over the two sets of values for ι, i.e.

$$y_p = \frac{1}{L} \sum_{m=0}^{P} x_m \cdot h_{p-m} + \frac{1}{L} \sum_{m=p+1}^{L-1} x_m \cdot h_{L+p-m} \qquad (6.27)$$

We saw earlier in Section 5.7.3 that this represents a 'circular' convolution in which a particular value obtained at time t is added to a value present at a time separated by $L/2$ sampling periods from t. To separate these we can consider h_ι to consist of the original filter series h_k, containing $M-1$ terms, plus $L-M$ zeros, so that for $L-1>p\geqslant M$ then the second summation in equation (6.27) will be zero and the convolution reduced to

$$y'_p = \frac{1}{L} \sum_{m=0}^{P} x_m \cdot h_{p-m} \qquad (6.28)$$

This will be seen to be equivalent to equation (6.17) for the smaller series of $L-M$ terms, taken from $N-1$ terms of the total series, if we neglect the scaling term $1/L$. To obtain this result we must therefore reject the first $M-1$ values of the series in y_p as productive of erroneous circular-convolved data.

This process is shown diagrammatically in Fig. 6.7. The lines indicate length of data sequences and terminate with an arrow followed by the

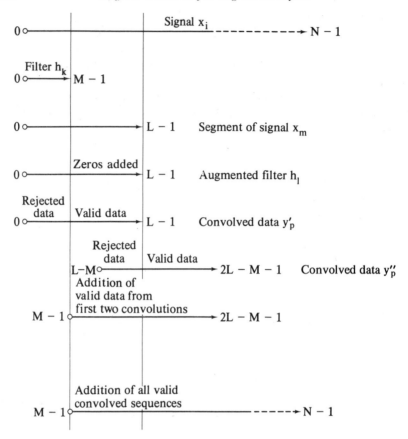

Fig. 6.7   The select-save method of continuous convolution

number of terms for that sequence. Thus the complete signal x_i to be filtered is shown as a line of length $N-1$ terms and the filter weights h_k by a much shorter line of $M-1$ terms. A section of the signal x_m is chosen of length $L-1$ terms and this is convolved with an augmented filter h_k of the same length, consisting of $M-1$ terms of the original filter plus $L-M$ zero terms.

The results of this convolution, y'_p, for the first section of the signal are seen to consist of $M-1$ terms of invalid data which are rejected, and the remaining $L-M$ terms which form the first acceptable fraction of the convolved data. The selected values for x_m forming the next section to be convolved, are taken from terms x_{L-M} to x_{2L-M-1} and a second convolution, y''_p, is carried out in the same way. After the rejection of the first $M-1$ terms this second fraction of the convolved data is added to the first as shown. Continuing in this way and adding the fractional convolutions along a time axis gives a total convolution series equivalent to that obtainable from the direct application of equation (6.17). The missing $M-1$ terms at the beginning of the signal are included by simply prefixing k zeros to the original data series, x_i. A flow diagram is given in Fig. 6.8 which indicates the various stages in the continuous convolution process. An optimum ratio for L/M given a suitable value for L is easily derived, since the choice is a compromise between a small number of lengthy convolutions or a larger number of smaller and hence faster convolutions.

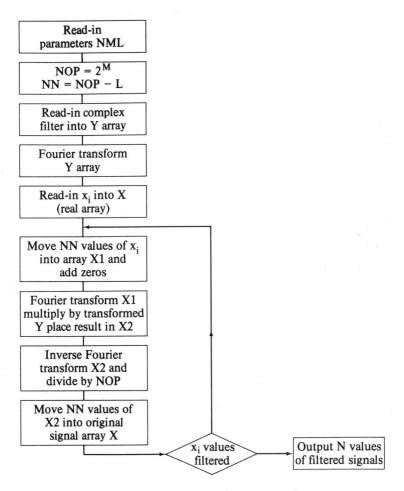

Fig. 6.8   A flow diagram for continuous convolution

The 'overlap-add' method is essentially similar. In this case $M-1$ zeros are added to the end of a section of the signal, x_m, so that when this is convolved with the filter the result contains no circularly convolved section. Consequently this may be overlapped by $M-1$ values with the result of the next partial convolution and added to it along the time axis. The computing times required for the two methods are very similar and there seems little to choose between them.

It will have been noted that a scaling factor, $1/L$, is included in the resultant convolution. As stated earlier the averaging term, $1/L$, for the FFT can appear in either the direct transform or the inverse transform. Since the process of obtaining a convolution by this method involves the inverse transformation of the product of two transforms, we can arrive at a situation in which the convolved value can be averaged by either $1/L$ or $1/L^2$, dependent on where we choose to include the averaging term. If this is included in the transform, as shown in equation (3.49), then the final output filter values will need to be multiplied by L to achieve a zero insertion loss.

6.2.2 CALCULATION OF FILTER WEIGHTS

Synthesis procedures for FIR filters can derive from any representation of the desired filter performance. If the required transfer function is known analytically, this can be directly transformed to the impulse response by way of the inverse Fourier transform. Alternatively the required frequency characteristics can be stated in graphical form, samples taken from this, and transformed to the time domain to form a series of filter weights. In either case an infinite series of non-zero Fourier series coefficients are required to represent the transformed frequency characteristic exactly. Since this is not possible an approximation must be obtained by truncation of the Fourier series. The nature of the approximations is well known and leads to the Gibbs phenomenon (Chapter 3) which manifests itself as a fixed percentage overshoot and ripple before and after the truncation discontinuity.

In a practical case a modified Fourier series is used in which a weighting function is applied on the raw Fourier series. This time-limited weighting function is called a **window**. A major requirement for this is that its defining impulse-response function must contain most of its energy in the main lobe, and have relatively small side lobes. A window used for digital filter work is that of a **cosine bell**, having the form

$$W_k = \frac{1}{2}\left(1 + \cos \frac{2\pi k}{N} \right) \quad (-\tfrac{1}{2}N \leqslant k \leqslant \tfrac{1}{2}N) \qquad (6.29)$$

This function contains 90 per cent of its energy in the main lobe and peak amplitude of the side lobes reduced to about 2 per cent of the fundamental peak. The effect of this in modifying a rectangular frequency function, (corresponding to truncation of a low-pass filter impulse-response function) is shown in Fig. 6.9. This improves the ripple in the pass-band at the expense of a slower fall-off at the band edges.

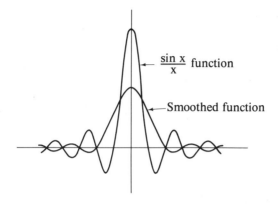

Fig. 6.9 Effect of smoothing the impulse-response function

The choice of a suitable smoothing window to minimise the effects of truncation of the Fourier transform was discussed in Chapter 4.

Before describing a general procedure for evaluating the filter weights it will be helpful to consider the equivalence between convolution and transformed products derived earlier. It has been shown (Section 6.1.2) that convolution in the time domain transforms to the products of transformation in the frequency domain, i.e.,

$$y(t) = h(t)*x(t) \Longleftrightarrow Y(\omega) = H(\omega).X(\omega) \qquad (6.30)$$

Where the quantities are expressed in discrete form the domain identification is lost, since we are manipulating a series of discrete numbers rather than a time- or frequency-dependent continuous function. Hence it is equally valid to state

$$y_i = h_k.x_i \Longleftrightarrow Y_i(\omega) = H_k(\omega)*Y_i(\omega) \qquad (6.31)$$

providing the equations are dimensionally correct.

A procedure for the derivation of the required filter weights via the FFT can be stated as follows:

1. The required response in the positive frequency domain is drawn and repeated in reversed order over the negative frequency domain (Fig. 6.10). The steepest discontinuity is noted and its frequency duration, f_d is divided into the Nyquist frequency ($f_s/2$). For an overshoot of less than 0.3 per cent peak in the achieved filter response the

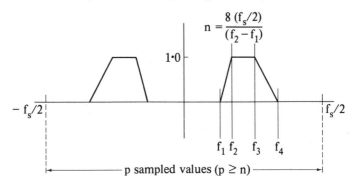

Fig. 6.10   Representation of desired frequency response

minimum number of filter weights, n, is given by Kuo and Kaiser [5] as

$$n \geqslant \frac{8(f_s/2)}{f_d} \qquad (6.32)$$

2.   Samples of the desired frequency response are taken at intervals of

$$\frac{f_s k}{p} \qquad \text{for } 0 \leqslant k < p/2$$

$$f_s \left(\frac{k}{p} - 1 \right) \text{ for } p/2 \leqslant k < p-1$$

where k = 0,1,...,p−1, and p⩾n   (6.33)

The amplitude values of the samples are taken as 1·0 within the pass-band to obtain correct scaling.

By sampling the desired frequency response at a number of frequencies, p, larger than the required number of filter weights, it is possible to obtain fairly good approximations to the first n Fourier coefficients.

3.   The samples are transformed, using an FFT routine and stored as a complex array.

4.   A smoothing window of n points is derived and multiplied point by point with the p transformed samples, such that the n significant products are retained and truncation to n values takes place.

5.   The series of n weights form the filter coefficient series, h_k, required for the filter.

To test the filter action and to compare this with the desired frequency response h_k can be transformed to form the filter transfer function using equation (6.19).

This procedure for obtaining filter weights is shown in Fig. 6.11. Note that the transformation from the required filter frequency response characteristic to the time domain can be obtained with either a DFT or IDFT since H_k is real. The choice of intermediate amplitude values between 0 and 1·0 in the frequency transition periods is only critical if the filter operates near the Nyquist limit. For many uncritical situations a linear set of amplitude values would be chosen, as indicated in Fig. 6.11.

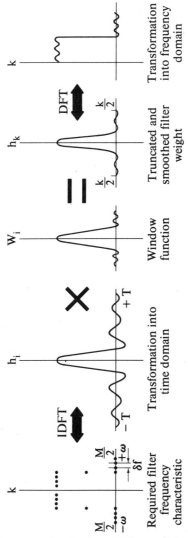

Fig. 6.11 Derivation of filter weights

An alternative method of filter definition is given by Ormsby [6] who found an improvement in filter performance for a given number of filter weights by introducing a specified value of slope into the filter characteristic (shown in Fig. 6.12 for the low-pass filter) where the transfer function is

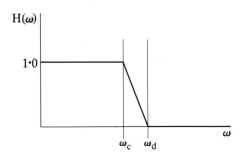

Fig. 6.12   The Ormsby low-pass filter shape

$$H(\omega) \begin{cases} = 1 & \text{for } 0 \leqslant \omega < \omega_c \\[2mm] = \left[\dfrac{\omega - \omega_d}{\omega_c - \omega_d} \right]^n & \text{for } \omega_c \leqslant \omega \leqslant \omega_d \qquad (6.34) \\[2mm] = 0 & \text{for } \omega_d \leqslant \omega \end{cases}$$

and n defines the rate of cut-off for the filter in units of 6 db/octave. The Ormsby filters are characterised by their ability to pass polynomials without change whilst reducing their noise content and, due to the symmetry, they have zero phase shift. A disadvantage is a slow fall-off in gain and the amount of ripple included in the pass-band of the realised filter.

For the low-pass filter it can be shown that a given set of filter weights can be applied for other cut-off frequencies and sampling intervals than the original designed values, providing we keep their product constant. Advantage may be taken of this by obtaining a long series of weights for a low value of f_c and small sampling interval, t, and using this as data for a simple decimation program. Filter weights for a given cut-off frequency, F_c, and sampling interval, T, are obtained by abstracting every R'th point (including the first value) where

$$R = \frac{F_c T}{f_c t} \qquad (R_{min} > R \geqslant 1) \qquad (6.35)$$

and R_{min} is a minimum value dependent on the attenuation characteristic and product $f_c t$ for the original filter, derived from equation (6.32).

It will be necessary to scale the abstracted weights to produce a filter having zero insertion loss. Thus if the input series, x_i, have unit value then from equation (6.17)

$$Y_i = \sum_{k=0}^{M-1} h_k = 1 \qquad (6.36)$$

Hence for a given (unscaled) series of filter weights, h_k, we can derive a new (scaled) series

$$h_k = \frac{h'k}{H} \qquad (6.37)$$

where H represents the summation of the unscaled series. This can be expressed simply by saying that for zero insertion loss a normalisation to unity of the area under the impulse-response function is required.

6.3 Digital Filtering with Walsh and Haar Series*

Digital filtering using non-sinusoidal orthogonal series has been carried out for both the one- and two-dimensional case. It has proved particularly valuable for operations involving very large data matrices (e.g. image enhancement) or real-time applications (e.g. television or radar processing). It has been noted earlier (Section 3.9.2) that arithmetic convolution cannot be applied to the non-sinusoidal series so that a relationship between transform products and convolution of two series such as derived in equation (6.14) is no longer valid. As a consequence the classical methods of direct Wiener filtering are used [7].

Considering the one-dimensional case, the input vector, x(t), is assumed to consist of additive zero-mean signal, s(t), and noise, n(t), uncorrelated with each other. Utilising a transformation operation, A consisting of a matrix of N by N values then

$$\begin{aligned} X(k) &= A.s(t) + A.n(t) \\ &= A(k) + N(k) \end{aligned} \qquad (6.38)$$

The resultant vector, X(k), is multiplied by an N by N matrix of filter weights, G_k, inversely transformed to produce a filtered output

$$y(t) = A^{-1}.G_kA.x(t) \qquad (6.39)$$

G_k chosen to provide the best mean-square estimate of the signal component, s(t), of the input vector, x(t) (Fig. 6.13).

In its simplest form for one-dimensional filtering G_k can consist of a string of ones followed by zeros corresponding to a total of N items in the transformed signal. Operation by equation (6.39) thus reduces the higher

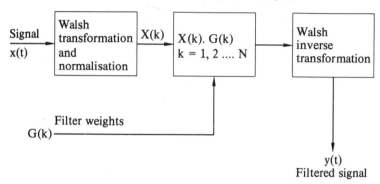

Fig. 6.13   Wiener filtering

sequency terms to zero, producing a low-pass sequency filter capable of being implemented with minimum hardware or which may be digitally processed at high speed.

Two possible types of filter matrices are applicable with this procedure. The filter matrix may be **vector** (i.e., contain non-zero components along the diagonal only) as with Fourier non-recursive filtering, or it may be **scalar** (containing non-zero components off the diagonal).

Vector filtering using a Walsh or Haar transformation can be extremely fast and may be carried out using $2N\log_2 N$ additions/subtractions plus N real multiplications. However very few practical filter requirements in terms of actual frequency-derived specifications result in a simple diagonal filter matrix.

Scalar filtering is the general case and although the process of transformation may be fast, multiplication by the filter matrix can demand up to N^2 multiplications. A reduction in computation time is obtained if many of the off-diagonal coefficients can be made zero and it is possible to design a sub-optimal filter of this type, using the Walsh series, from known properties of a Fourier filter which is economical in both design and computation time [8].

If we express the two forms of filtering in matrix terms then we can write for Fourier filtering the output column vector $y_1(t)$ as

$$y_1(t) = \mathbf{F}^{-1}\mathbf{G}_1\mathbf{F}.x(t) \qquad (6.40)$$

where $\mathbf{F}$ and $\mathbf{F}^{-1}$ are the direct and inverse Fourier transform, and $\mathbf{G}_1$ is the set of filter weights.

Walsh filtering can be written in the same form as

$$y_2(t) = \mathbf{W}^{-1}\mathbf{G}_2\mathbf{W}.x(t) \qquad (6.41)$$

where $\mathbf{W}$ and $\mathbf{W}^{-1}$ are the direct and inverse Walsh transform, and $\mathbf{G}_2$ is a second set of filter weights appropriate to Walsh filtering. If we assume similar outputs for the two filtering operations then, $y_1(t) \simeq y_2(t)$ and we can write

$$\mathbf{F}^{-1}\mathbf{G}_1\mathbf{F}.x(t) = \mathbf{W}^{-1}\mathbf{G}_2\mathbf{W}.x(t) \qquad (6.42)$$

from which

$$\mathbf{G}_2 = \mathbf{W}.\mathbf{F}^{-1}\mathbf{G}_1\mathbf{F}\mathbf{W}^{-1} \qquad (6.43)$$

The filter weights, $\mathbf{G}_1$, may be derived using well-tried methods. Using the fast Fourier and Walsh transforms, the required filter weights, $\mathbf{G}_2$, necessary for Walsh vector filtering are obtained. Whilst this process of filter weight derivation will be slower then simple determination of $\mathbf{G}_1$ for Fourier filtering, the actual productive process of filtering using the Walsh transform can be considerably faster. This would be important for complex or repetitive filtering operations such as two-dimensional image filtering, or on-line television applications.

6.4 The IIR Digital Filter

A disadvantage of the FIR filter is that a large number of filter weights is necessary to approximate the desired function. We may, for example, require 200 filter weights to obtain a satisfactory approximation to the ideal shape for a given filter shown in Fig. 6.5. This means 200 multiplications/additions per data point which can lead to lengthy computational time in addition to demanding substantial memory storage.

IIR filters will reduce both of these requirements by an order of magnitude. They will, however, be found to exhibit a more complex structure in which the output includes the weighted sum of both input and output terms. Due to the economy in filter series length a fast acting filter is obtained which makes it valuable in real-time applications. Whilst implementation of the IIR filter can be carried out using the fast Fourier transform [9] this does not confer such great advantages as with the FIR case and the following discussion is limited to direct evaluation of the difference equation. The IIR filters described below will, in general, have **non-zero** phase characteristics. In many applications this is unimportant but where this is likely to affect adversely the transient response to the data then the following procedure can be adopted. A filter is designed to give half the attenuation slope required. The data are passed through the filter twice, once in the usual way, and a second time with the data points arranged in reverse order. The phase changes will be found to cancel and a real value of transfer function obtained. A slight correction will be required for cut-off frequency.

A linear difference equation for the IIR filter was given as equation (6.3), and is known as the compound filter. Its transfer function may be derived in a similar manner to that described for the filter as

$$H(\omega) = \frac{Y(\omega)}{X(\omega)} = \frac{\displaystyle\sum_{k=0}^{M} a_k \cdot \exp(-j\omega kT)}{1 - \displaystyle\sum_{k=0}^{P} b_k \cdot \exp(-j\omega kT)} \tag{6.44}$$

which has been shown by Holz and Leondes [10] to be a rational function of sines and cosines having a polynomial form, and therefore equivalent to the general form for the continuous filter given by equation (6.1).

A simpler form of the IIR filter is given as

$$y_i = Cx_i + \sum_{k=0}^{P} b_k \cdot y_{i-k} \tag{6.45}$$

for which its Fourier transform yields a transfer function

$$H(\omega) = \frac{Y(\omega)}{X(\omega)} = \frac{C}{1 - \displaystyle\sum_{k=0}^{P} b_k \cdot \exp(-j\omega kT)} \tag{6.46}$$

This type of relation is somewhat easier to implement since the poles of the transfer function are sufficient to define the filter weights.

A simple example is the first-order difference low-pass filter given by

$$y_i = x_i + Ky_{i-1} \quad (i = 1,2,\ldots,N) \tag{6.47}$$

The transfer function can be obtained by letting $y_i = A \sin i\omega t$, where $\omega = 2\pi f$ and $f = 1/2T$. Thus

$$y_i = KA \sin i\omega(t - T) + x_i$$

$$= KA \sin i\omega t . \cos i\omega T - KA \cos i\omega t . \sin i\omega T + x_i$$

But $\sin i\omega T = \sin i(2\pi T/2T) = \sin i\pi = 0$, since i is an integer. Therefore $y_i = Ky_i . \cos i\omega T + x_i$ and

$$H(f) = y_i/x_i = \frac{1}{1 - (K \cos i\omega T)} \tag{6.48}$$

K can be related to filter cut-off frequency defining f_c to be at the half-power point

$$| H(f) |^2 = \frac{1}{2} = \frac{1}{1 + (K^2 \cos^2 i\omega T) - (2K \cos i\omega T)}$$

but $\cos^2 i\omega T = \cos^2[i(2\pi T/2T)] = \cos i\pi = 1$ since i is an integer, so that

$$K^2 - 2K \cos \omega_c T = 1$$

where $\omega_c = i\omega$ and $K = \cos \omega_c T \pm \sqrt{(\cos^2 \omega_c T + 1)}$. This can be substituted in the recursive expression given in equation (6.47) to enable a simple algorithm to be implemented for the low-pass filter. Note that only one multiplication and one addition is required to realise a single output filter point.

A number of design techniques have been developed for the IIR filter and whilst some methods enable the design to be carried out completely in the frequency domain [11], a considerable simplification ensues if the z-transform is used. A brief introduction to the z-transform is given in the next section.

6.4.1 THE Z-TRANSFORM

One way of defining the z-transform is from the Laplace transform and, since filter theory for continuous filters is expressed in Laplace form, we can consider the z-transform as a logical extension of this for a discrete series. A sampled series, x_i, can be considered to be the product of a continuous signal, $x(t)$, and a set of uniformly spaced unit impulses. Thus the Laplace transform, $H(s)$, of a sampled series can be expressed as

$$H(s) = \sum_0^\infty [a_0(t) + a_1\delta(t - T) + a_2\delta(t - 2T) + \ldots,] \exp(-sT)$$

(6.49)

where $a_0, a_1, a_2, \ldots$, represent the sample amplitudes and T is the sampling interval. Hence

$$H(s) = \sum_{k=0}^\infty a_k \exp(-ksT) \tag{6.50}$$

This may be expressed in simplified terms to facilitate the algebraic manipulation by replacing $\exp(sT)$ by z, thus

$$\exp(sT) = z, \text{ or}$$
$$\exp(-sT) = z^{-1} \tag{6.51}$$

and by replacing $H(s)$ by $H(z)$ so that

$$H(z) = \sum_{k=0}^\infty a_k z^{-k} \tag{6.52}$$

$H(z)$ is thus by definition the z-transform of x_i and represents a power series in z^{-1} with coefficients, a_k representing the amplitude of successive samples of x_i in the time domain.

Tables of z-transforms for sampled data series can be obtained similar to the Laplace transforms derived for a continuous signal by using the transformation given by equation (6.51). It is convenient to use the transform for z^{-1} rather than z since the multiplicative factor of z^{-1} is equivalent to a delay of T, or a delay of one sampling period.

Some properties of the z-transform will now be derived by consideration of a number of typical functions for x_i.

(i) Step function: $x_i = \begin{cases} 1 : n \geqslant 0 \\ 0 : n < 0 \end{cases}$

From equation (6.52)

$$H(z) = z^0 + z^{-1} + z^{-2}, \ldots, z^{-n}$$

If $z^{-1} < 1$ then the series converges to

$$H(z) = \frac{1}{1 - z^{-1}} \qquad (6.53)$$

(ii) $x_i = a^n$

$$H(z) = \sum_{n=0}^{\infty} a^n z^{-n} = \sum_{n=0}^{\infty} (az^{-1})^n$$

which converges to

$$H(z) = \frac{1}{1 - az^{-1}} \qquad (6.54)$$

(iii) A sampled data series: $x_i = x[nT] = \exp(-anT)$

$$H(z) = \sum_{n=0}^{\infty} \exp(-anT)z^{-n} = \frac{1}{1 - \exp(-aT)z^{-1}} \qquad (6.55)$$

(iv) $x_i = na^n$

$$H(z) = \sum_{n=0}^{\infty} na^n z^{-n} = z \sum_{n=0}^{\infty} na^n z^{-(n+1)}$$

$$= z \sum_{n=0}^{\infty} a^n \frac{dz^{-n}}{dz}$$

$$H(z) = z \left[\frac{d}{dz} \sum_{n=0}^{\infty} a^n n^{-n} \right]$$

so that substituting H(z) from equation (6.54) and differentiating

$$H(z) = \frac{az}{(1 - az^{-1})^2} \qquad (6.56)$$

from this example we can obtain the general relationship

$$H[(z)nx(t)] = z \left[\frac{dH(z)[x(t)]}{dz} \right] \qquad (6.57)$$

(v) A sampled data series delayed by K units: $x_i = x[(n - k)T]$

$$H(z) = \sum_{n=0}^{\infty} z[(n - k)T]z^{-n} = z^{-k} \sum_{n=0}^{\infty} x[(n - k)T]z^{-(n-k)}$$

which, from equation (6.52)

$$H(z) = z[x(n - k)T] = z^{-k}[x(nT)] \qquad (6.58)$$

This is the **delay property** or shifting theorem of the z-theorem. Each delay by one sampling interval corresponds to a multiplication by z^{-1} in the z-domain.

Thus $H(z) = z^{-2}$ corresponds to a sample taken with a delay of two sampling units, i.e. x_{i-2}. Similarly $H(z) = z^3$ corresponds to a forward shift in time by three sampling intervals, i.e., x_{i+3}.

In general the z-transform possesses equivalent properties to those of the Laplace transform. The z-transfer function, H(z), is equal to the ratio of the output and input functions expressed as z-transforms. Multiplication of functions by z is equivalent to convolution of functions of time. Finally a complex function can be expressed graphically on a z-plane in much the same way as it can be described in terms of the s-plane. The two representations are related in a manner which will be discussed in Section 6.4.2.

An important application of the z-transform which we shall be using in this chapter is in the representation of difference equations. Continuous time functions can be regarded as the solutions of differential equations in which integration, represented as 1/s in the s-domain, plays an essential part. Similarly, for sampled time functions, these are the solutions of difference equations in which the essential element is that of unit delay, represented by z^{-1} in the z-domain.

A transfer function for the recursive filter was given in equation (6.44).

This can be expressed in z-form as

$$H(z) = \frac{\displaystyle\sum_{k=0}^{M} a_k z^{-k}}{1 - \displaystyle\sum_{k=0}^{P} b_k z^{-k}} \tag{6.59}$$

The filter is realisable when the filter parameters, a_k and b_k, are real and constant and where the poles of $H(z)$ lie inside the unit circle on the z-plane (see later). Equation (6.59) can also be expressed in product form as a general function of z corresponding to equation (6.2)

$$H(z) = \frac{Y(z)}{X(z)} = K \frac{(z - a_1)(z - a_2),\ldots,(z - a_n)}{(z - b_1)(z - b_2),\ldots,(z - b_m)} \tag{6.60}$$

where $a_1, a_2, \ldots, a_n$ are zeros, and $b_1, b_2, \ldots, b_m$ are simple poles.

The polynomials in z represented by $Y(z)$ and $X(z)$ have real coefficients since a_n and b_m are all either real or occur in complex conjugate pairs. Since a multiplication in z implies a time shift of one sampling interval the z-transfer function can be converted easily into a difference equation. An example will be given using a transfer function having one zero and two complex poles, viz.

$$H(z) = \frac{Y(z)}{X(z)} = \frac{z - a}{(z - \alpha - j\beta)(z - \alpha + j\beta)} \tag{6.61}$$

where a is a zero in the z-plane, and $\alpha \pm j\beta = b_1, b_2$ are complex poles in the z-plane.

This can be rearranged as

$$Y(z)[z^2 - 2\alpha z + \alpha^2 + \beta^2] = X(z)[z - a]$$

and divided by z^2 to give

$$Y(z)[1 - 2\alpha z^{-1} + (\alpha^2 + \beta^2)z^{-2}] = X(z)[z^{-1} - az^{-2}] \tag{6.62}$$

Replacing the present input and output sampled values by x_i and y_i and invoking the shifting theorem then equation (6.62) can be written

$$y_i - 2\alpha y_{i-1} + (\alpha^2 + \beta^2)y_{i-2} = x_{i-1} - ax_{i-2}$$

or

$$y_i = x_{i-1} - ax_{i-2} + 2\alpha y_{i-1} - (\alpha^2 + \beta^2)y_{i-2} \tag{6.63}$$

This recurrence relationship or difference equation allows a new value for the current output value, y_i, to be obtained from past input and past output values, weighted by given z-plane zero and pole coefficients.

6.4.2. MAPPING THE S-PLANE INTO THE Z-PLANE

The Laplace transform of a single pole continuous function, $x(t) = a_i \exp (s_i t)$ is by definition

$$X(s) = \frac{a_i}{(s - s_i)} \qquad (6.64)$$

A sampled data series, $x_i = x(nT)$, may be formed from this continuous series, $x(t)$, by sampling at uniform intervals, separated by T. This can be regarded as a delta series modulated by an amplitude function which, in this case, is an exponential term. Thus

$$x(nT) = \sum_{n=0}^{\infty} a_i \exp (s_i t)\delta(t - nT)$$

$$= a_i[\delta(t) + \exp (s_i t)\delta(t - T) +,\ldots,+ \exp (s_i t)\delta(t - nT)] \qquad (6.65)$$

This represents a series of delayed impulses in which the coefficients, $a_i \exp (s_i T)$, $a_i \exp (2s_i T)$ etc., are constants.

The Laplace transform of a unit impulse is given as $\exp (-snT)$, so that the transform of the sampled data series may be represented as

$$X(s) = \sum_{n=0}^{\infty} a_i \exp (s_i - s)T = a_i[1 + \exp (s_i - s)T + \exp (s_i - s)2T +,\ldots,$$

$$+ \exp (s_i - s)nT] \qquad (6.66)$$

which is a geometrical progression and converges to

$$X(s) = \frac{a_i}{1 - \exp (s_i - s)T} \qquad (6.67)$$

This may be compared with the Laplace transforms of a continuous function given by equation (6.64).

The poles of equation (6.67) occur where $\exp (s_i - s)T = 1$. Thus $(s_i - s)T = 0$ and $s = s_i$ which correspond to the single pole of the continuous function. However we can also find unit value for $\exp (s_i - s)T$ when this is equal to $\exp (\pm j2\pi n)$ where $n = 0,1,2,\ldots$, so that: $(s_i - s)T = \pm j2\pi n$, and

$$s = s_i \pm j\frac{2\pi n}{T} = s_i + jn\omega_s \qquad (6.68)$$

where ω_s is the sampling frequency, $2\pi f_s$.

Hence the effect of sampling on a continuous form of a transfer function, $H(s)$, will take the form of folding or repetition of the frequency characteristic so that instead of $H(s)$ we obtain a sampled version

$$H^*(s) = \sum_{n=-\infty}^{\infty} H(s + jn\omega_s) \qquad (6.69)$$

which is equivalent to

$$H(z) = T \sum_{n=0}^{\infty} h(nT)z^{-n} \qquad (6.70)$$

Thus the pole-zero pattern of the original signal will be repeated at intervals of $2\pi/T$ over the s-plane shown in Fig. 6.14. This is a result of the sampling theory discussed earlier and has similar consequences. It is essential to ensure that the sampling rate for the data to be filtered is adequate to match the particular filter design if stability is to be preserved.

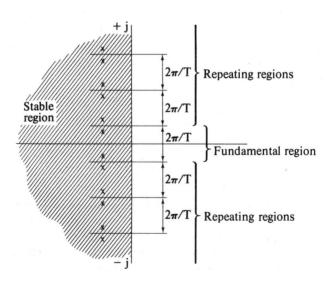

Fig. 6.14   Repeated poles of a recursive filter

For wide-band filters where $\omega_c \simeq \omega_s/2$ then folding errors can occur and a solution proposed by Kuo and Kaiser [5] is to precede the desired wide-band filter, $H(s)$, by a low-pass filter, $G(s)$, having a high attenuation slope to give a cascaded filter

$$H_c(s) = H(s) \cdot G(s) \qquad (6.71)$$

This may be transformed into z-transfer form and implemented. The resultant z-transfer function, $H_c(z)$, will however be complex and contain considerably more terms than the simpler realisation for $H(s)$.

The correspondence between the s-plane and the z-plane can be seen if we write

$$z = \exp(sT) = \exp(\alpha + j\omega)T \qquad (6.72)$$

so that

$$|z| = \exp(\alpha T) \text{ and } <z = \omega T$$

and a point s_i in the s-plane will transform to a point z_i in the z-plane (Fig. 6.15).

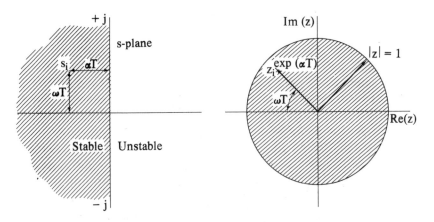

Fig. 6.15   Relationship between the s-plane and the z-plane

Since the path along the imaginary axis for the s-plane where the poles are situated corresponds to $\alpha = 0$ then $|z| = 1$ and the angle $<z$ varies between $\pm\pi$ radians. We thus infer that the $j\omega$ axis in the s-plane maps into a unit radius circle in the z-plane. The left-hand half of the s-plane maps inside the unit circle of the z-plane. This is the region of stability for a system where the uncancelled poles must lie within the unit circle. The effect of containing the entire stable region of the s-plane within the unit circle of the z-plane means that all the repeated poles of the sampled function will now result in a single pole within the unit circle in the z-plane.

This has the disadvantage that undesirable aliasing is not readily apparent from the z-plane representation.

The z-transform offers considerable manipulational convenience due to the elimination of exponential terms and normalisation with respect to the sampling interval. It also possesses the ability to directly translate a polynomial in z to a difference equation by the use of the shifting theorem. In the next section we will consider synthesis methods using this s- to z-plane transformation.

6.4.3 GENERAL METHODS

Synthesis methods for recursive filters operate either from considerations of the pole-zero requirements or indirectly from the transfer function of the desired continuous filter.

An example of the first of these is the direct synthesis from the impulse-transfer function into a z-transform realisation and its implementation as a difference equation or set of difference equations derived from this realisation.

Three examples of the indirect method are:

1. Frequency domain synthesis from squared magnitude transfer functions. This is similar to a method of design for continuous filters, in which the poles and zeros are first determined to define the filter characteristics. From these the filter weights are obtained directly and the recursive expression, given by equation (6.3), is implemented.
2. Frequency domain synthesis from the continuous filter considered in the s-plane and known as the bi-linear transform. This method overcomes the folding disadvantage of the standard z-transform by mapping the entire complex plane into a single horizontal strip in the s-plane corresponding to the fundamental region shown in Fig. 6.14. It is an indirect method in which this band-limited s-plane realisation is converted into the z-plane and from this a difference equation derived which can be implemented as a recursive algorithm.
3. Simulation of the filter characteristics in the frequency domain by means of a frequency sampling technique [12]. Here the process of sampling is used to reconstruct the filtered signal from its elemental time responses.

A brief discussion of these methods is given below and is followed by a fuller discussion of one of these, namely the bi-linear transform method, which is attractive due to its simple realisation and economy.

The direct synthesis from the impulse-transfer function is a method of design from the known characteristics of continuous systems in terms of its poles and zeros, and is sometimes referred to as the **impulse-invariance** method. The transfer function is given and a sampled version of the impulse-response function obtained, which can be converted into a difference equation and used to simulate the recursive digital filter [5]. This method will be explained in terms of a single real pole transfer function

$$H(s) = \frac{1}{s+a} \qquad (6.73)$$

where the single pole at $s = -a$ is known. The Laplace transform of a discrete series of unit impulses $x_i = x(nT)$ has been shown from equation (6.67) for a pole $s_i = -a$ to be

$$X(s) = \frac{k}{1-\exp(-a-s)T} \qquad (6.74)$$

or

$$H(z) = \frac{k}{1-\exp(-aT)z^{-1}} \qquad (6.75)$$

The filter can be realised from the recurrence relation

$$Y_i = kx_i + \exp(-aT)Y_{i-1} \qquad (6.76)$$

To determine k we equate the gains of the continuous and discrete transfer function at zero frequency. Thus with $s = j\omega = 0$ equation (6.73) becomes $1/a$ and equation (6.75) becomes $k/[1-\exp(-aT)]$, so that

$$k = \frac{1-\exp(-aT)}{a} \qquad (6.77)$$

To use this method the continuous transfer function is factorised into one or more elementary transfer functions such as that given by equation (6.73). Each function is converted into its corresponding discrete version and the polynomial in z converted into a difference equation using the shifting theorem. The programming of the recursive algorithm is simplified if the exponential term, $\exp(-aT)$, is calculated outside the recursive loop since this only needs to be calculated once after the sampling rate has been determined. The impulse method has been applied to the calculation of equiripple Lerner filters referred to in Section 4.6.4 for the filter-bank method of spectral evaluation.

The FIR filters and a number of the IIR filters discussed in the previous sections are characterised by symmetrical weighting functions and zero phase-shift. Their characteristics in other directions are however far from optimum. A class of filters having symmetrical weighting functions can be defined by a transfer function having the form

$$H(s) = \frac{1}{\displaystyle\prod_{i=1}^{m} (s_i + s)} \qquad (6.78)$$

These filters have m poles at finite values of z and zeros at infinity. They can be designed to have optimum amplitude characteristics but will exhibit pronounced phase shift in the region of cut-off frequency.

The most important of these are the **Butterworth** and **Chebychev** filters which are defined by the following squared-magnitude functions

$$|H(j\omega)|^2 = \frac{1}{1 + \left(\dfrac{\omega}{\omega_c}\right)^{2n}} \qquad \text{(Butterworth)} \qquad (6.79)$$

and

$$|H(j\omega)|^2 = \frac{1}{1 + E^2 V n^2 \left(\dfrac{\omega}{\omega_c}\right)} \qquad (6.80)$$

The pole positions for a Butterworth filter will be found to lie equally spaced around a circle of radius ω_c in the s-plane and confined to the left-hand half. It may also be shown [13] that the poles of a Chebychev filter are arranged in an ellipse whose major axis lies along the imaginary axis in the s-plane. Figure 6.16 shows the pole locations for both of these filter types.

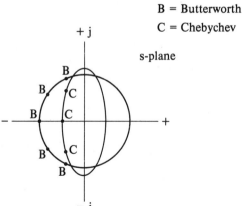

B = Butterworth
C = Chebychev

s-plane

Fig. 6.16   Pole-zero representation of a Butterworth filter

The transfer function can be obtained by consideration of the pole locations. Thus for a second-order Butterworth filter (n = 2)

$$H(s) = \frac{1}{1+P_a} \cdot \frac{1}{1+P_b} \tag{6.81}$$

The poles are located at $\pi/4$ radius with respect to the real axis. Their coordinates are

$$\left. \begin{array}{l} -\omega \cos \pi/4 + j\omega \sin \pi/4 \text{ for pole A} \\ \omega \cos \pi/4 + j\omega \sin \pi/4 \text{ for pole B} \end{array} \right\} \tag{6.82}$$

Hence, substituting in (6.81)

$$H(s) = \frac{1}{\left(1 - \dfrac{2^{1/2}\omega}{2} + j\dfrac{2^{1/2}\omega}{2} \right) \left(1 + \dfrac{2^{1/2}\omega}{2} + \dfrac{2^{1/2}\omega}{2} \right)}$$

$$= \frac{1}{1 + 2^{1/2}j\omega + (j\omega)^2}$$

and in normalised form, referred to cut-off frequency ω_c, with $j\omega$ replaced by s

$$H(s) = \frac{1}{1 + 2^{1/2}(s/\omega_c) + (s/\omega_c)^2} \tag{6.83}$$

which is the required transfer function.

A digital transfer function closely resembling that of the Butterworth low-pass filter has been proposed by Holz and Leondes [10], viz.

$$|H(j\omega)|^2 = \frac{1}{1 + \left[\dfrac{\tan^{1/2}\omega T}{\tan^{1/2}\omega_c T} \right]^{2n}} \tag{6.84}$$

where ω_c is the cut-off frequency, defined as the frequency at which the amplitude squared value for $H(\omega)$ is reduced by half from its value at $\omega = 0$ (−3db). A similar derivation for the Chebychev filter gives

$$|H(j\omega)|^2 = \frac{1}{1 + E^2 V_n^2 \left[\dfrac{\tan^{1/2}\omega T}{\tan^{1/2}\omega_c T} \right]} \tag{6.85}$$

where V_n is the n'th order Chebychev polynomial and E is related to the fractional pass-band ripple.

The phase characteristics for both filters are non-linear but their amplitude characteristics are optimum for a given number of filter weights. A better attenuation characteristic outside the pass-band is achieved by the Chebychev filter at the expense of a controlled amount of constant amplitude ripple in the pass-band given by equation (6.85). Both Butterworth and Chebychev filters exhibit rather lengthy settling time (for a second-order Butterworth low-pass filter this can be several periods at the cut-off frequency). An improvement can be obtained with elliptic filters at the expense of a finite ripple content in both pass-band and stop-band [2].

Equation (6.83) can be re-written in z-transform notation by letting $z = \exp(j\omega T)$ and substituting

$$\tan(^1/_2\omega T) = j\, \frac{\exp(j^1/_2\omega T) - \exp(-j^1/_2\omega T)}{\exp(j^1/_2\omega T) + \exp(-j^1/_2\omega T)}$$

so that

$$|H(z)|^2 = \frac{k^n}{k^n + (-1)^n \left[\dfrac{z-1}{z+1}\right]^{2n}} \qquad (6.86)$$

where $k = \tan^2(^1/_2\omega_c T)$.

The poles of the function are found by substituting $s = \frac{z-1}{z+1}$ in equation (6.86) and will be found to be uniformly spaced around a circle of radius, $\tan(^1/_2\omega_c T)$ in the s-plane, which thus corresponds with that found for equation (6.79). The poles for the Chebychev filter will be found to lie on an ellipse in the s-plane and can be determined by a similar substitution of $s = \frac{z-1}{z+1}$.

The recursive realisation of these filters can be obtained by multiplying out the numerator and denominator polynomials in H(z) and applying the shifting theorem. An alternative realisation has been proposed by Otnes [14] where the filter weights are determined by conformal mapping, retaining the results in terms of a polynomial of complex frequency terms. In this case the coefficients for the recursive equation are evaluated in terms of a natural frequency and damping ratio. A suitable computational procedure for this is described by Enochson and Otnes [11] for a number of filter types. These realisations can lead to difficulties in implementation on the digital computer if high-order filters are attempted. This will be discussed further in Section 6.4.6.

The effects of sampling in the time domain have been considered earlier. If a continuous signal, band-limited to $\pm B$ H_z, is sampled, the original signal can be reconstructed by passing the sampled signal through a filter having a sin x/x shape for its impulse-response function. Since each sampled value can be considered as exciting the filter to produce a sin x/x response we can visualise the simulation of the complete filter response as comprising the summation of each individual sample response in the frequency domain (Fig. 6.17). This principle is used in the frequency sampling filter [12]. The signal is applied to a comb filter which is designed to have m zeros equally spaced around the unit circle. The filter is followed by a lossless resonator, defined by two complex conjugate poles which lie directly on the unit circle. If the angle of the resonator pole is made equal to one of the zeros of the comb filter then an output at the fundamental resonant frequency of the resonator is obtained. If this pole-zero cancellation is not obtained then little output results. By applying the comb filter output to m resonators connected in parallel, each resonant at a different discrete frequency related to the 'teeth' frequencies of the comb, then the output response of each resonant filter can be added as a scalar quantity.

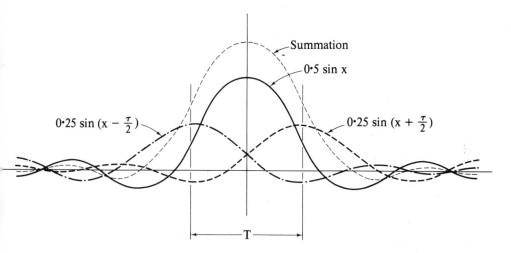

Fig. 6.17   Synthesis method used in frequency sampling filters

This is shown in Fig. 6.18. The impulse response of the resonators is arranged to have a finite period, mT, to limit its contribution, considered in the time domain, following excitation. The resonators are followed by attenuators set to provide a gain at each resonator frequency corresponding to the required filter response. Summation of the attenuated resonator responses gives a synthesised output corresponding to the filtered signal.

The designed filter has a linear phase characteristic and allows the specification of response in the frequency domain as the attenuation required at each frequency sampled value. However, since m−1

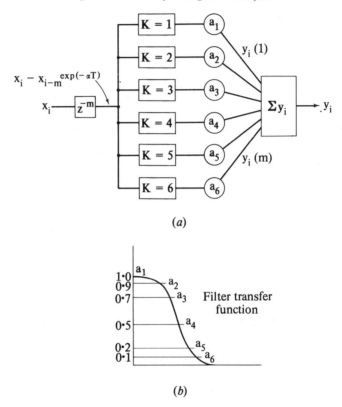

(a)

(b)

Fig. 6.18   Method of frequency sampling

resonators are required then a substantial amount of memory is required in the computer to retain the intermediate results before attenuation and addition are carried out to form the final output. These storage requirements approach those needed for FIR filters. Realisation of the comb filters response is obtained from the difference equation

$$y_i = x_{i-m} \cdot \exp(-\alpha T) \qquad (6.87)$$

which requires simply a series of m delays. The lossless sinusoidal resonator is represented as

$$y_i = x_i + y_{i-1}[\exp(j\omega k - \alpha)T] \qquad (6.88)$$

where $k = 1, 2, \ldots, m - 1$.

From the preceding brief description the filter synthesis will be seen to have similarities with Fourier synthesis methods and shares a common requirement with them of requiring much greater digital storage than the simpler recursive techniques.

6.4.4 THE BI-LINEAR Z-TRANSFORM*

The two advantages of this method are:
1. The transformation is purely algebraic in form and so is applicable to continuous analog transfer functions.
2. Aliasing errors, possible with the direct z-transform method, are removed.

It is an indirect method since the requirements of an equivalent continuous filter are first postulated and then transformed into a discrete form, from which the filter weights can be derived. Essentially the transformation is from the integrating operation $1/s$ to the z-transform.

If we consider a continuous signal, $x(t)$, to be operated upon by an ideal integrator having a transfer function, $H(s) = 1/s$, then the output, $y(t)$, can be represented by $x(t)*h(t)$ where * represents a convolution operation with the impulse-response function for $H(s)$. An integrator having a transfer function $H(s) = 1/s$ is of course ideal. Its impulse-response function will be a step function; $h(t) = 1$ for $t \geqslant 0$; and $h(t) = 0$ for $t < 0$ and we can define the results of such integration on a signal, $x(t)$, as the convolution operation

$$y(\tau) = \int_0^t x(\tau) . h(t - \tau)d\tau \qquad (6.89)$$

since $h(t)$ becomes zero for $\tau > t$ and unity for $\tau \leqslant t$. This can be expressed in discrete form by considering t_1 and t_2 to be two consecutive sampling times separated by the sampling interval T where

$$x(t) = x(nT) \text{ and } n = 0,1,2,\ldots,$$

so that the continuous integral defined by equation (6.88) and having limits t_1 and t_2 approximates to

$$\tfrac{1}{2}T[x(nT) + x(n - 1)T] = y(nT) - y(n - 1)T$$

and we can write for the z-transfer function

$$H(z) = \frac{y(z)}{x(z)} = \tfrac{1}{2}T \left[\frac{1 + z^{-1}}{1 - z^{-1}} \right] \qquad (6.90)$$

This represents a trapezoidal approximation to integration using the s-operator expressed as a function of z. Thus the transfer function, $H(s)$ for continuous system can be replaced by the z-transfer function $H(z)$ using the relationships

$$\frac{1}{s} \rightarrow \tfrac{1}{2}T \left[\frac{z + 1}{z - 1} \right]$$

$$s \rightarrow \frac{2}{T}\left[\frac{z-1}{z+1}\right] \qquad (6.91)$$

giving

$$H(z) \equiv H(s) \Bigg|_{s=\frac{2}{T}\left[\frac{z-1}{z+1}\right]} \qquad (6.92)$$

This transform is known as the **bilinear z-transform** [15]. It maps the imaginary axis of the s-plane into a unit circle of the z-plane such that the left-hand side of the s-plane corresponds to the interior of the circle (Fig. 6.15). Thus repeating poles of a discrete transfer function resulting from aliasing effects are coalesced into single unique positions within the circle (assuming a stable filter) removing the possibility for aliasing errors present with the direct z-transform method [16]. If the transformation is carried out precisely as a unique 1:1 relationship the filter will be stable and of the same order as the original Laplace transfer function. The frequencies of the two relationships will, however, be different so that the replacement of s shown in equation (6.92) will result in a digital filter having a cut-off angular frequency, ω_d, related to the continuous frequency ω_a, by

$$\omega_a = \frac{2}{T} \tan\left[\frac{\omega_d T}{2}\right] \qquad (6.93)$$

This has the effect of compressing the complete continuous frequency characteristics into a limited digital filter frequency range of $0 < \omega T < \pi$ as indicated in Fig. 6.19. In the practical application of equations (6.92) and (6.93) we will be calculating the ratio s/ω_a so that the 2/T term may be dropped and we can write

$$s = \frac{z-1}{z+1} \qquad (6.94)$$

$$\omega_a = \tan\left[{}^1\!/_2\omega_d T\right] \qquad (6.95)$$

This is the form in which these terms often appear in the literature.

The method of design using the bilinear transform can be summarised as follows:

1. A new set of frequencies for the continuous transfer function, H(s), is calculated from the desired digital frequencies using equation (6.95).
2. A suitable continuous transfer function, H(s), is chosen to give the required filter performance using the frequencies derived from step 1.

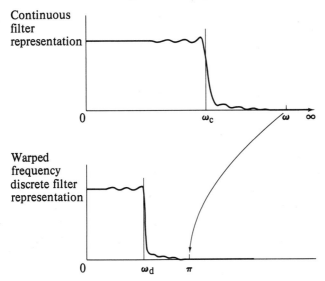

Fig. 6.19   Warping of the frequency scale

3.   The operator s in H(s) is replaced by a function in z and the new transfer function, H(z), expressed as a ratio of polynomials in z.
4.   H(z) is converted into a difference equation by the application of the shifting theorem and the algebraic equation rearranged to give a recursive equation for the output sample.
5.   The recursive equation, considered as an algorithm, is programmed for the digital computer.

This procedure will now be described for a number of the filter types.

A low-pass Butterworth filter will be taken as the filter model. The transfer functions for the first-order (n = 1) and second-order (n = 2) have been given previously as

$$H(s) = \frac{1}{1 + s/\omega_c} \qquad (6.96)$$

and

$$H(s) = \frac{1}{1 + 2^{1/2}(s/\omega_c) + (s/\omega_c)^2} \qquad (6.97)$$

These transfer functions will be used as a basis for design, commencing with the low-pass case.

(i) Low-pass filter design

A difference equation will be developed for a first-order (n = 1) and a second-order (n = 2) transfer function, using equations (6.96), (6.97), (6.94) and (6.95).

1st-order case. Given a cut-off frequency, f_c Hz, the required equivalent continuous cut-off angular frequency is from (6.95).

$$\omega_a = \tan(\pi f_c T) \qquad (6.98)$$

The z-transform function is from (6.96) and (6.94)

$$H(z) = \frac{Y(z)}{X(z)} = \frac{1}{1 + \left[\dfrac{z-1}{z+1} \quad \dfrac{1}{\omega_a}\right]}$$

$$= \frac{z+1}{z\left[1 + \dfrac{1}{\omega_a}\right] + \left[1 - \dfrac{1}{\omega_a}\right]}$$

Multiplying by z^{-1} and equating input and output terms

$$Y(z)\left[\left(1 + \frac{1}{\omega_a}\right) + z^{-1}\left(1 - \frac{1}{\omega_a}\right)\right] = X(z)[z^{-1} + 1]$$

$$(6.99)$$

and applying the shifting theorem gives

$$y_i = Gx_i + Gx_{i-1} - Hy_{i-1} \qquad (6.100)$$

where

$$G = \frac{1}{1 + \cot(\pi f_c T)}$$

and

$$H = \frac{1 - \cot(\pi f_c T)}{1 + \cot(\pi f_c T)}$$

The filter gain is A = 1/G.

2nd-order case. The z-transfer function is from (6.97) and (6.94)

$$H(z) = \frac{1}{1 + \left[\dfrac{2^{1/2}}{\omega_a} \cdot \dfrac{z-1}{z+1}\right] + \left[\dfrac{z-1}{z+1}\right]^2 \cdot \dfrac{1}{\omega_a^2}}$$

Substituting for ω_a from equation (6.95) and invoking the shifting theorem, we can write the difference equation for the second-order case as

$$y_i = Cx_i + 2Cx_{i-1} + Cx_{i-2} - Dy_{i-1} - Ey_{i-2} \qquad (6.101)$$

where

$$C = \frac{1}{1 + [2^{1/2} \cdot \cot(\pi f_c T)] + [\cot(\pi f_c T)]^2}$$

$$D = 2[1 - [\cot(\pi f_c T)]^2]$$

and

$$E = 1 - [2^{1/2} \cdot \cot(\pi f_c T)] + [\cot(\pi f_c T)]^2$$

The filter gain is $1/C$.

Transformation of the continuous low-pass filter transfer functions to the equivalent high-pass, band-pass and band-stop forms can be obtained through well-known conformal mapping techniques [17]. A list of these is given in Table 6.1. Use is made of the low-pass/high-pass transformation as described below.

Table 6.1   **s-plane transformations of low-pass continuous filters**

Required filter	Replace s/ω_c by:
Low-pass	s/ω_c
High-pass	ω_c/s
Band-pass	$\dfrac{s^2 + \omega_l\omega_u}{s(\omega_u - \omega_l)}$
Band-stop	$\dfrac{s(\omega_u - \omega_l)}{s^2 + \omega_l\omega_u}$

ω_c = cut-off angular frequency,
ω_l = lower cut-off angular frequency,
ω_u = upper cut-off angular frequency.

For the band-pass and band-stop transformation more economical algorithms are obtained if the transformation is carried out directly in the z-plane and these will be used in the following derivations.

(ii) High-pass filter design
1st-order case. From Table 6.1 and equations (6.96), (6.94) and (6.95), the z-transfer function is

$$H(z) = \frac{z - 1}{z(1 + \omega_a) + (\omega_a - 1)}$$

from which the difference equation is obtained as

$$y_i - G' x_i - G' x_{i-1} - H' y_{i-1} \qquad (6.102)$$

where

$$G' = \frac{1}{1 + \tan(\pi f_c T)}$$

and

$$H' = \frac{1 - \tan(\pi f_c T)}{1 + \tan(\pi f_c T)}$$

The filter gain is $1/G'$

2nd-order case. The z-transform function is obtained as

$$H(z) = \frac{(z - 1)^2}{(z - 1)^2 + 2^{1/2}\omega_a(z^2 - 1) + \omega_a^2(z + 1)^2}$$

from which the difference equation is

$$y_i = C' x_i - 2C' x_{i-1} + C' x_{i-2} + D' y_{i-1} - E' y_{i-2} \qquad (6.103)$$

where

$$C' = \frac{1}{1 + 2^{1/2}\tan(\pi f_c T) + [\tan(\pi f_c T)]^2}$$

$$D' = 2| 1 - (\tan(\pi f_c T))^2]$$

and

$$E' = 1 - 2^{1/2}(\tan(\pi f_c T) + [\tan(\pi f_c T)]^2$$

The filter gain is $1/C'$

We can note from the similarity of equations (6.100) and (6.102), also (6.101) and (6.103) that only two algorithms are needed to evaluate first- and second-order low-pass and high-pass filters. To convert from a low-pass to a high-pass difference equation we need only to

(*a*) change the sign of the odd-numbered weights for both x_i and y_i,
(*b*) substitute $\tan(\omega f_c T)$ for $\cot(\omega f_c T)$ in the evaluation of the filter weights.

It can also be shown that if only the signs of the weights are changed for the low-pass filter then it will behave as a high-pass filter having a cut-off frequency of $((1/2T) - f_c)$.

(iii) Band-pass filter design

The transformation from the s-plane to a cylindrical surface resulting from the extension of the unit circle in an amplitude domain has been described by Constantinides [18]. This results in a new transformation which goes directly from the Laplace transfer function of a low-pass filter to the discrete band-pass case given by

$$s = \frac{z^2 - 2z\alpha + 1}{z^2 - 1} \tag{6.104}$$

where

$$\alpha = \frac{\cos \pi T(f_u + f_l)}{\cos \pi T(f_u - f_l)} = \cos \omega_0 T \tag{6.105}$$

ω_0 represents the warped band centre frequency which is derived from the difference of the warped values of f_u and f_l and is different from

$$\omega_a = \tan \pi T(f_u - f_l) \tag{6.106}$$

which represents the equivalent mean frequency of the continuous filter and will not be central.

These relationships will now be used to derive difference equations for band-pass filter.

1st-order case. Substituting (6.104) in (6.96) gives the z-transfer function

$$H(z) = \frac{1}{1 + \dfrac{(z^2 - 2z\alpha + 1)}{z^2 - 1} \cdot 1/\omega_a}$$

Carrying out algebraic manipulation and dividing by z^2 gives

$$H(z) = \frac{\omega_a - \omega_a(z^{-2})}{(\omega_a + 1) - 2\alpha z^{-1} + (1 - \omega_a)z^{-2}} = \frac{Y(z)}{X(z)}$$

from which the difference equation may be obtained as

$$y_i = Jx_i - Jx_{i-2} + Ky_{i-1} + Ly_{i-2} \tag{6.107}$$

where

$$J = \frac{1}{1 + 1/\omega_a} = \frac{1}{1 + \cot \pi T(f_u - f_l)}$$

$$K = \frac{2\alpha}{\omega_a + 1} = \frac{2 \cos \pi T(f_u + f_l)}{[1 + \tan \pi T(f_u - f_l)] \cos \pi T(f_u - f_l)}$$

and

$$L = \frac{(\omega_a - 1)}{(\omega_a + 1)} = \frac{\tan \pi T(f_u - f_l) - 1}{\tan \pi T(f_u - f_l) + 1}$$

The filter gain is $1/J$.

2nd-order case. Similarly for the second-order case

$$H(z) = \left[\frac{1 + 2^{1/2}}{\omega_a} \left(\frac{z^2 - 2z\alpha + 1}{z^2 - 1} \right) + \left(\frac{z^2 - 2z\alpha + 1}{\omega_a(z^2 - 1)} \right)^2 \right]^{-1}$$

After algebraic manipulation and dividing by z^4 we obtain

$$H(z) = \frac{\omega_a^2(z^{-4} - 2z^{-2} + 1)}{(\omega_a^2 + 2^{1/2}\omega_a + 1) + z^{-1}(-2 \cdot 2^{1/2}\omega_a\alpha - 4\alpha) + z^{-2}(-2\omega_a^2 + 4\alpha^2 + 2) + z^{-3}(2 \cdot 2^{1/2}\omega_a\alpha - 4\alpha) + z^{-4}(\omega_a^2 - 2^{1/2}\omega_a + 1)}$$

from which the difference equation can be obtained as

$$y_i = Mx_i - 2Mx_{i-2} + Mx_{i-4} - Oy_{i-1} - Py_{i-2} - Qy_{i-3} - Ry_{i-4} \tag{6.108}$$

where $M = \omega_a^2/N$, $N = \omega_a^2 + 2^{1/2}\omega_a + 1$, $O = (-2 \cdot 2^{1/2}\omega_a\alpha - 4\alpha)/N$, $P = (-2\omega_a^2 + 4\alpha^2 + 2)/N$, $Q = (2 \cdot 2^{1/2}\omega_a\alpha - 4\alpha)/N$, $R = (\omega_a^2 - 2^{1/2}\omega_a + 1)/N$, and ω_a and α are defined by equations (6.105) and (6.106). The filter gain is $1/M$.

(iv) Band-stop filter design
Similar equations apply for the band-stop filter. The warped analog frequency, ω_a, is now:

$$\omega_a = \cot \pi T(f_u - f_l) \tag{6.109}$$

and the transformation relationship is

$$s = \frac{z^2 - 1}{z^2 - 2z\alpha + 1} \tag{6.110}$$

1st-order case. From equations (6.96) and (6.110).

$$H(z) = \cfrac{1}{1 + \left(\cfrac{z^2 - 1}{z^2 - 2z\alpha + 1}\right) \cdot \cfrac{1}{\omega_a}}$$

Rearranging and dividing by z^2 gives:

$$H(z) = \frac{\omega_a - 2z^{-1}.\alpha\omega_a + \omega_a.z^{-2}}{(\omega_a + 1) - z^{-1}.2\alpha\omega_a + z^{-2}(\omega_a - 1)} = \frac{Y(z)}{X(z)}$$

from which the difference equation can be obtained as

$$y_i = J'x_i - K'x_{i-1} + J'x_{i-2} + K'y_{i-1} - L'y_{i-2} \tag{6.111}$$

where

$$J' = \frac{1}{1 + 1/\omega_a} = \frac{1}{1 + \tan \pi T.(f_u - f_l)}$$

$$K' = \frac{2\alpha\omega_a}{\omega_a + 1} = \frac{2\cos\pi T.(f_u - f_l)\cot\pi T.(f_u - f_l)}{\cot(f_u - f_l).(T + 1)}$$

and

$$L' = \frac{\omega_a - 1}{\omega_a + 1} = \frac{\cot\pi T.(f_u - f_l) - 1}{\cot\pi T.(f_u - f_l) + 1}$$

The filter gain is $1/J'$.

2nd-order case. For the second-order case

$$H(z) = \left[1 + \frac{2^{1/2}}{\omega_a}\left(\frac{z^2 - 1}{z^2 - 2z\alpha + 1}\right) + \frac{1}{\omega_a^2}\cdot\left(\frac{z^2 - 1}{z^2 - 2z\alpha + 1}\right)^2 \right]$$

Rearranging and dividing by z^4 gives

$$H(z) = \frac{\omega_a^2(1 - 4z^{-1}.\alpha + z^{-2}(4\alpha^2 + 2) - 4z^{-3}.\alpha + 2z^{-4})}{(\omega_a^2 + 2^{1/2}\omega_a + 1) - z^{-1}(4\alpha\omega_a^2 + 2\cdot2^{1/2}\omega_a\alpha) + z^{-2}(4\alpha^2\omega_a^2 + 2\omega_a^2 - 2)}$$
$$+ z^{-3}(-4\alpha\omega_a^2 + 2\cdot2^{1/2}\alpha\omega_a) + z^{-4}(\omega_a^2 - 2^{1/2}\omega_a + 1)$$

from which the difference equation can be obtained as

$$y_i = M'x_i - Sx_{i-1} + Tx_{i-2} - Sx_{i-3} + M'x_{i-4}$$
$$+ Oy_{i-1} - P'y_{i-2} - Q'y_{i-3} - R'y_{i-4} \tag{6.112}$$

where: $M' = \omega_a^2/N'$, $N' = \omega_a^2 + 2^{1/2}\omega_a + 1$, $O' = (4\alpha^2 . \omega_a^2 + 2 \cdot 2^{1/2}\omega_a\alpha)N'$, $P' = (4\alpha^2 . \omega_a^2 + 2\omega_a^2 - 2)/N'$, $Q' = (-4\alpha . \omega_a^2 + 2 \cdot 2^{1/2}\omega_a\alpha)/N'$, $R' = (\omega_a^2 - 2^{1/2}\omega_a + 1)/N'$, $S = 4\alpha\omega_a^2/N'$, $T = \omega_a^2(4\alpha^2 + 2)/N'$, and ω_a and α are defined by equations (6.105) and (6.109). The filter gain is $1/M'$.

The transformations and resulting difference equations for these filter designs are summarised in Tables 6.2 and 6.3.

6.4.5 ERRORS IN DIGITAL FILTER REALISATION

A number of errors are associated with a given design of digital filter. Performance errors such as phase or amplitude error have already been considered. Errors peculiar to the use of sampled data are:

1. Quantisation of the input data.
2. Quantisation of the calculated values for the filter weights.
3. Quantisation of the results of iterative mathematic operations (i.e. round-off and truncation errors).
4. Aliasing errors.
5. Errors introduced due to the dynamic range of the input data (e.g. effects of overflow in the accumulator).

Quantisation of the input data has been considered earlier where it was shown to be equivalent to introducing an additional noise signal at the filter input. The average noise power varies with the filter realisation and becomes less as more zeros are introduced into the polynomial transfer

Table 6.2 s-Plane to z-plane transformation from a low-pass continuous filter

	s is replaced by:	ω_a is replaced by:
Low-pass/Low-pass	$\dfrac{z - 1}{z + 1}$	$\tan(\pi f_c T)$
Low-pass/High-pass	$\dfrac{z + 1}{z - 1}$	$\cot(\pi f_c T)$
Low-pass/Band-pass	$\dfrac{z^2 - 2z\alpha + 1}{z^2 - 1}$	$\tan \pi T(f_u - f_l)$
Low-pass/Band-stop	$\dfrac{z^2 - 1}{z^2 - 2z\alpha + 1}$	$\cot \pi T(f_u - f_l)$.

T = Sampling interval, f_c = cut-off frequency, f_l = lower cut-off frequency.
f_u = upper cut-off frequency, $\alpha = [\cos \pi T(f_u + f_l)]/[\cos \pi T(f_u - f_l)]$.

Table 6.3 Summary of filter difference equations derived from a continuous Butterworth low-pass filter design

Type	Order (n)	Equation	Coefficients
Low-pass	1	$y_i = Gx_i + Gx_{i-1} - Hy_{i-1}$	$G = [1 + \cot(\pi f_c T)]^{-1}$, $H = [1 - \cot(\pi f_c T)]/[1 + \cot(\pi f_c T)]$
	2	$y_i = Cx_i + 2Cx_{i-1} + Cx_{i-2} - Dy_{i-1} - Ey_{i-2}$	$C = [1 + 2^{1/2}\cot(\pi f_c T) + [\cot(\pi f_c T)]^2]^{-1}$
			$D = 2[1 - [\cot(\pi f_c T)]^2]$
			$E = (1 - 2^{1/2}\cot(\pi f_c T) + [\cot(\pi f_c T)]^2)$
High-pass	1	$y_i = G'x_i - G'x_{i-1} - H'y_{i-1}$	$G' = [1 + \tan(\pi f_c T)]^{-1}$, $H = [1 - \tan(\pi f_c T)]/[1 + \tan(\pi f_c T)]$
	2	$y_i = C'x_i - 2C'x_{i-1} + C'x_{i-2}$ $\qquad + D'y_{i-1} - E'y_{i-2}$	$C' = [1 + 2^{1/2}\tan(\pi f_c T) + [\tan(\pi f_c T)]^2]^{-1}$
			$D' = 2[1 - [\tan(\pi f_c T)]^2]$
			$E' = 1 - 2^{1/2}\tan(\pi f_c T) + [\tan(\pi f_c T)]^2$
Band-pass	1	$y_i = Jx_i - Jx_{i-2} + Ky_{i-1} + Ly_{i-2}$	$J = [1 + \cot \pi T(f_u - f_l)]^{-1}$
			$K = [2 \cos \pi T(f_u - f_l)]/[1 + \tan \pi T(f_u - f_l) \cdot \cos \pi T(f_u - f_l)]$
			$L = [\tan \pi T(f_u - f_l) - 1]/[\tan \pi T(f_u - f_l) + 1]$
	2	$y_i = Mx_i - 2Mx_{i-2} + Mx_{i-4}$ $\qquad - Oy_{i-1} - Py_{i-2} - Qy_{i-3} - Ry_{i-4}$	$M = [\tan^2 \pi T(f_u - f_l)]/N$
			$N = \tan^2 \pi T(f_u - f_l) + 2^{1/2}\tan \pi T(f_u - f_l) + 1$
			$O = [-\cos \pi T(f_u + f_l)] [2 \cdot 2^{1/2}\tan T(f_u - f_l) + 4]/[\cos \pi T(f_u - f_l)]N$
			$P = [-2\tan^2\pi T(f_u - f_l) + 2 \cdot 2^{1/2}\tan \pi T(f_u - f_l) + \left(\dfrac{2 \cos \pi T(f_u + f_l)}{\cos \pi T(f_u - f_l)} \right)^2 + 2]/N$
			$Q = \cos \pi T(f_u + f_l)[2 \cdot 2^{1/2}\tan \pi T(f_u - f_l) - 4]/[\cos \pi T(f_u - f_l)]N$
			$R = [\tan^2\pi T(f_u - f_l) - 2^{1/2}\tan \pi T(f_u - f_l) + 1]/N$

Table 6.3 (Contd.)

Type	Order (n)	Equation	Coefficients
Band-stop	1	$y_i = J' x_i - K' x_{i-1} + J' x_{i-2} + K' y_{i-1} - L' y_{i-2}$	$J' = [1 + \tan \pi T(f_u - f_l)]^{-1}$
			$K' = 2 \cos \pi T(f_u - f_l) \cot \pi T(f_u - f_l)/[\cot \pi T(f_u - f_l) + 1]$
			$L' = [\cot \pi T(f_u - f_l) - 1]/[\cot \pi T(f_u - f_l) + 1]$
	2	$y_i = M' x_i - S x_{i-1} + T x_{i-2} - S x_{i-3}$ $+ M' x_{i-4} + O' y_{i-1} - P' y_{i-2}$ $- Q' y_{i-3} - R' y_{i-4}$	$M' = [\cot^2 \pi T(f_u - f_l)]/N'$
			$N' = \cot^2 \pi T(f_u - f_l) + 2^{1/2} \cot \pi T(f_u - f_l) + 1$
			$O' = [\cos \pi T(f_u + f_l)][2 \cdot 2^{1/2} \cot \pi T(f_u - f_l) + 4 \cot^2 \pi T(f_u - f_l) \cot \pi T(f_u - f_l)]/[\cos \pi T(f_u - f_l)]N'$

$$P' = \left[\, 2 \cot^2 \pi T(f_u - f_l) + \left(\frac{2 \cos \pi T(f_u + f_l) \cot \pi T(f_u - f_l)}{\cos \pi T(f_u - f_l)} \right)^2 - 2 \,\right] / N'$$

$$Q' = [\cos \pi T(f_u + f_l)][2 \cdot 2^{1/2} \cot \pi T(f_u - f_l) - 4 \cot^2 \pi T(f_u - f_l)]/(\cos \pi T(f_u - f_l)]N$$

$$R = [\cot^2 \pi T(f_u - f_l) - 2^{1/2} \cot \pi T(f_u - f_l) + 1]/N$$

$$S = [4 \cos \pi T(f_u - f_l) \cdot \cot^2 \pi T(f_u - f_l)]/\cos \pi T(f_u - f_l)N'$$

$$T = \cot^2 \pi T(f_u - f_l) \left[\, \left(2\, \frac{\cos \pi T(f_u + f_l)}{\cos \pi T(f_u - f_l)} \right)^2 + 2 \,\right] / N'$$

T = sampling interval, f_c = cut-off frequency, f_u = upper cut-off frequency, f_l = lower cut-off frequency

function. A similar effect obtains from a consideration of the quantisation of the products and sums obtained at each iteration of the recursive equation. The cumulative effect of round-off error in digital filter calculation depends on the realisation of the filter transfer function. It will be shown in the next section that the minimal noise contribution is obtained with cascade representation.

Quantisation of the filter weights is necessary due to the finite word length of the computer. The effect of inaccuracy in the quantised values may be seen from the following example. If we consider a transfer function

$$H(z) = \frac{1}{(1 - a_1 z^{-1})(1 - a_2 z^{-1})} \qquad (6.113)$$

having two poles at distances $1/a_1$ and $1/a_2$ removed from the origin of the z-plane, then defining the desired values of the poles as A_1 and A_2 the partial derivatives against the coefficient values, a_1 and a_2 for each of the cascaded realisations

$$H_1(z) = \frac{1}{1 - a_1 z^{-1}} \text{ and } H_2(z) = \frac{1}{1 - a_2 z^{-1}}$$

will be

$$\frac{\partial A_1}{\partial a_1} = 1, \qquad \frac{\partial A_2}{\partial a_2} = 1$$

$$\frac{\partial A_1}{\partial a_2} = \frac{\partial A_2}{\partial a_1} = 0$$

If equation (6.113) is now realised in direct form as

$$H(z) = \frac{1}{1 - b_1 z^{-1} + b_2 z^{-1}} \qquad (6.114)$$

where $b_1 = a_1 + a_2$ and $b_2 = a_1 . a_2$, we can state the partial derivatives of the derived poles A_1 and A_2 against actual coefficients b_1 and b_2 as

$$\frac{\partial A_1}{\partial b_1} = 1 \qquad \frac{\partial A_2}{\partial b_2} = \frac{1}{a_1}$$

$$\frac{\partial A_1}{\partial b_2} = \frac{1}{a_2} \qquad \frac{\partial A_2}{\partial b_1} = 1$$

This shows that

(a) change in the filter coefficients will produce a change in the pole positions,
(b) a change in the direct form coefficients will be productive of a larger pole position change than the cascaded form coefficients.

In some cases the inaccuracy in coefficient value can cause the pole positions to move outside the unit circle thus causing instability. As a general rule it is inadvisable to implement filters higher than the second order in direct form.

The digital computer will set a limit to the magnitude of numbers that can be represented. Overflow error can occur during the summing and product operations and will generally show itself as a repeated oscillation at the output. This effect will be more severe with fixed point arithmetic as compared with floating point operation and a limit may need to be imposed at the input of the filter to avoid these effects.

6.4.6 IMPLEMENTATION OF M'TH-ORDER FILTER

Difficulties can be experienced when the coefficients of high-order IIR filters are determined. In particular with a very large change in gain at the pass-band edges, then a wide range in filter weight coefficients will be realised. For computers having a small word length it will not be possible to represent this range adequately since the larger coefficients will be specified to a greater degree of accuracy than the smaller ones. In larger machines having floating-point hardware then double-precision working can be used to retain the dynamic range required.

The effects of this **dynamic error**, which is represented by a quantisation of the desired filter coefficients, has been described in the previous section in terms of pole position on the z-plane. It has been shown [19] that if the poles are located within 10^{-n} of the unit circle, the filter weights will need to be defined with an accuracy of n decimal places to achieve stability. Even within this stability margin a satisfactory performance may not be achieved, due to these coefficients' inaccuracies.

A better solution is to reduce the filter order and achieve the desired performance by means of **cascaded transfer functions**

$$H(z) = H_1(z) \cdot H_2(z) \cdot H_3(z), \ldots, H_n(z) \qquad (6.115)$$

An accurate stable filter can be designed for each of the fractional transforms so that the composite filter can be considered as comprising a number of smaller filters connected in series with the output of one filter forming the input data for the next.

Only two filter algorithms need be considered. A first-order (single real pole) and a second-order (complex pole pair) filter from which any combination can be obtained. The method has been generalised by Knowles and Edwards [20] and termed cascade programming. The z-transform of any complex filter can be expressed as

$$H(z) = A \prod_{i-1}^{n} \frac{(1 - a_i z^{-1})(1 - a^*_i z^{-1})}{(1 - b_i z^{-1})(1 - b^*_i z^{-1})} \tag{6.116}$$

where A is the filter gain and * indicates a complex conjugate (since in a realisable filter any complex poles or zeros must exist in conjugate pairs).

Realisation of equation (6.115) consists of the cascade evaluation of simple polynomials having the form

$$H_i(z) = \frac{1 - a_i z^{-1}}{1 - b_i z^{-1}} \tag{6.117}$$

or their conjugate equivalent, with a suitable gain factor interposed between each product term. In addition to the reduced value of quantisation noise resulting from the calculation of coefficient values in cascading operation the cumulative error due to round-off will also be reduced by the method of cascading.

6.4.7 PROGRAMMING A CASCADE FILTER

As indicated earlier only two forms of the elementary filter need be considered. Suitable first-order and second-order difference equations were developed in Section 6.4.4. Specification of the four filter types can be defined by selection of two frequency values f_l and f_u, located at the band edges, thus for

Low-pass: $f_l = 0, f_u = f_c$
High-pass: $f_l = f_c, f_u = 0$
Band-pass: $f_l = f_{cl}, f_u = f_{cu}$
Band-stop: $f_l = f_{cu}, f_u = f_{cl}$

where f_c = cut-off frequency, f_{cl} = lowest cut-off frequency, and f_{cu} = higher cut-off frequency. A sign logic test can then act as a check on selected filter type.

Due to the close similarity between the low-pass and high-pass equations a single algorithm can be evolved to calculate either case. To obtain the desired cut-off slope, defined in decibels/octave, the order of the composite filter is calculated from equation (6.79) and hence the number of elemental filters required. Thus for a slope of S db/octave then

$$S = 10 \log_{10}(P_i/P_0)$$

where

$$P_i/P_0 = \frac{1}{|A|^2} = 1 + (\omega/\omega_c)^{2n}$$

so that for $(\omega/\omega_c) = 2$ (i.e., one octave)

$$S = 10 \log_{10} (1 + 2^{2n})$$

giving

$$n = 1 \cdot 662 \log_{10} (10^{0 \cdot 1S} - 1) \qquad (6.118)$$

The obtained value of n is rounded up to the next integer value and tested. If n is odd then a single first-order stage is required and $\frac{1}{2}(n - 1)$ second-order stages. If n is even then only $(n/2)$ second-order stages are required.

References

1. JURY, E. I. *Theory and Application of the z-Transform Method*. John Wiley, New York, 1964.
2. RADER, C. M., and GOLD, B. *Digital Processing of Signals*. McGraw-Hill, New York, 1969.
3. STOCKHAM, T. G. High-speed convolution and correlation. *AFIPS Proc. Spring Joint Comp. Conf.*, **28**, 229–33, 1966.
4. HELMS, H. D. Fast Fourier transform method of computing difference equations and simulating filters. *IEEE Trans. Audio and Electroacoust.*, **AU-15**, 85–90, June 1967.
5. KUO, F. F., and KAISER, J. F. *System Analysis by Digital Computer* (Chapter 7). John Wiley, New York, 1966.
6. ORMSBY, J. F. A. Design of numerical filters with applications to missile data processing, *JACM*, **8**, 3, 440–66. July 1961.
7. PRATT, W. K. Generalized Wiener filtering computational techniques. *IEEE Trans. Comp.*, **C.21**, 7, 636, 41, 1972.
8. KAHVECI, A. E., and HALL, E. L. Frequency domain design of sequence filters. *1972 Proceedings: Applications of Walsh Functions*. Washington, D.C. AD.744650, 1972.
9. GOLD, B. and JORDAN, K. A note on digital filter synthesis. *Proc. IEEE*. **56**, 1717–18, Oct. 1968.
10. HOLZ, H., and LEONDES, C. T. The synthesis of recursive digital filters. *JACM*, **13**, 2, 262–8, April 1966.
11. ENOCHSON, K. D., and OTNES, R. K. Programming and analysis for digital time series data. Shock and Vibration Information Center, U.S. Department Defence, 1968.
12. BOGNOR, R. E. Frequency sampling filters, Hilbert transformers and resonators. *Bell. Syst. Tech. J.*, **3**, 501, March 1969.
13. CONNOR, F. R. *Networks*. Arnold, London, 1972.

14. OTNES, R. K. An elementary design procedure for digital filters. *IEEE Trans. Audio and Electroacoust.*, **AU16**, 3, 330–5, Sept. 1968.
15. KAISER, J. F. Design methods for sampled-data filters. *Proc. First Allerton Conf. on circuit and system theory.* November 1963.
16. KAISER, J. F. Some practical considerations in the realisation of linear digital filters. *Proc. Third Allerton Conf. on circuit and system theory.* 621–33, Oct. 1965.
17. GUILLEMIN, E. A. *Synthesis of Passive Networks*. John Wiley, New York, 1957.
18. CONSTANTINIDES, A. G. Spectral transformations for digital filters. *Proc. IEE*, **117**, 8, 1585–9, August 1970.
19. KNOWLES, J. B., and OLCAYNO, E. M. Coefficient accuracy and digital filter response. *Electronic Letters*, **1**, 6, 160–1, August 1965.
20. KNOWLES, J. B., and EDWARDS, R. Effects of a finite word length computer in a sampled data feedback system. *Proc. IEEE*, **12**, 6, June 1965.

Chapter 7

Two-Dimensional Processing

7.1 Introduction

The previous chapters have been concerned with processing of single-dimensional series which may be expressed in terms of a single time-dependent parameter. Of increasing importance in recent years is the acquisition and processing of pictorial data where the information is defined by two spatial characteristics and some form of intensity level value corresponding to the single parameter found in one-dimensional series. We will call these data **image data**, and their behaviour with time expressed by an **image series**.

The most noticeable feature of this form of data is quantity which generally exceeds by several orders of magnitude time series data available from experimental equipment. This is reflected in the techniques used for computer image processing where the emphasis is on speed and efficiency and a modular approach to data handling. This has given rise to special developments in parallel processing and optical methods which are applicable for particular installations handling large amounts of similar data, e.g. processing of transmitted satellite images.

In this chapter we will consider some of the processing techniques developed for general-purpose high-speed digital computers where it is now feasible to form mathematical and algorithmic processes on images of natural photographic quality. These images are currently being collected from a wide spread of operations. Examples are found in aerial reconnaissance, including earth resource investigations, cloud patterns from weather satellites, X-ray photographs and those derived from nuclear research equipment, chromosome and neuron micro-photographs and many others [1, 2]. Automatic data analysis methods are essential either to render the large masses of data acquired to a form suitable for further analysis or simply to improve the subjective value of these images.

Image processing includes all the various operations which can be applied to data of this kind, e.g. image compression and restoration, image enhancement, pre-processing, transformation and quantisation of data and spatial filtering.

Although some authorities include pattern recognition as an image processing operation the subject is considered to be too specialised for inclusion in this book.

The object of image processing is in most cases to remove errors and to improve the image for subjective evaluation and further analysis by the investigator. In many cases the techniques used are simply extensions of the one-dimensional processes discussed earlier. The terminology is often different and new approaches are necessary to enable the large amount of data to be handled in general-purpose computers through a segmentation approach. Before considering these techniques it is necessary to discuss some of the basic definitions used in image processing.

7.2   Basic Definitions in Image Processing

7.2.1   SAMPLING AND QUANTISATION

A major problem of sampling and quantisation is to preserve the pictorial information upon reconstruction without deteriorating the value of the original information. The value of the reconstructed image is essentially subjective and will depend, in the first instance, on the way in which the initial digitisation is carried out and the method of synthesis, quite apart from the processing enhancements involved. A method of sampling and quantisation suitable for one particular image and application may be subjectively inadequate in another context.

In most cases sampling of the image area is carried out by first dividing the image area into sub-areas (often known as **pixels** or **picture elements**). These sub-areas are represented by one sample point which is made equal to the average grey level of the sub-area or to the grey level at the actual point of sampling. Whilst sub-areas are usually taken at a regular matrix spacing there may be advantages in non-constant spacing to correct, for example, perspective errors introduced by the collection device, e.g. camera tilt.

Synthesis of the sampled picture can be obtained by linear interpolation of the grey levels between the sampled points. Other forms of interpolation can be used. If sinusoidal functions are used for interpolation purposes then the sampling theorem holds and under certain conditions the picture can be reconstructed exactly [3]. For a two-dimensional image $f(x,y)$ having the maximum spatial frequencies of f_1 for x and f_2 for y, then sampling frequencies of $fs_1 \geqslant 2f_1$ for x and $fs_2 \geqslant 2f_2$ are required. Usually a unique value of f_s is chosen equal to the greater of fs_1 and fs_2 and a sampling interval of $h = 1/fs$.

Quantisation of the sampled data refers to assigning a particular grey level for the data sample taken. The samples can assume only a prespecified and finite set of grey levels, replacing the actual grey level at each point by the grey level nearest to it. If q is the quantisation step then the image grey level range is represented through Nq discrete grey levels where N is a suitable integer. Again, a set of evenly-spaced grey levels are generally used but there may be advantages in an unequal-spacing of the levels on a logarithmic basis in order to improve the dynamic range over a portion of the grey spectrum. A pseudo-random quantisation has been proposed to remove an undesirable outline to the picture content. A varying rate of quantisation has also been employed to take advantage of the observer's visual limitations which demand fine quantisation in smooth areas of the image and a coarser quantisation near the edges of the image [4].

If we represent the original image as a real-valued function of two variables, $f(x,y)$, expressed on a plane domain given in terms of Cartesian coordinates then the digitised image may be defined as a series of real numbers $x_{i,k}$. These can be expressed in matrix form as $[A_k]$ where A_k is a digital picture element or pixel and $k = 1,2,\ldots,N$.

Digitisation is the construction of a sequence of real numbers, $[A_k]$, from image information, $f(x,y)$, such that the image information is preserved within a specified error criterion. Mathematically the process can be defined as

$$A_k = F_k \left[f(x,y)[P_l] \right] \qquad (7.1)$$

where (P_l) is a finite parameter set and F_k is a transformation procedure.

Similarly the reconstruction of a picture from the digitised sequence (A_k) can be obtained through a further transformation G_k to obtain a two-dimensional image, $g(x,y)$, viz.

$$g(x,y) = G_k[[A_k]x,y] \qquad (7.2)$$

The error may be defined in terms of a comparison method as

$$f(x,y) \; \copyright \; g(x,y) \leqslant \epsilon(x,y) \qquad (7.3)$$

If the comparison method $\copyright$ is a mean-square error one then

$$\iint [f(x,y) - g(x,y)]^2 dx dy \leqslant \epsilon \qquad (7.4)$$

To carry out this digitisation process the image must be sampled and the resulting samples quantised into a finite number of bits. The process can be analysed by extension from the one-dimensional signal processing described in Chapter 1.

Since an image is not usually band-limited, aliasing can be expected in sampling an image. We saw earlier that in the case of one-dimensional processing aliasing is avoided by passing the signal through a low-pass filter prior to digitisation. An equivalent operation in the case of image processing is to adjust the aperture dimensions in the digitisation device and to select the sampling frequency accordingly. Aliasing of images is most apparent when the visual structures are periodic. This results in Moiré patterns in the reconstructed image which can only be avoided by increasing the sampling rate.

Many systems used to digitise images are modifications of known equipment used for microdensitometry [5]. Here the image is prepared as a transparent film and a small spot of light transmitted through the film or reflected from it and collected in the detection device producing an electrical signal proportional to the intensity of the light received. If the amount of light, I_2, transmitted through the film is compared with the amount of light, I_1, observed in the absence of the film the transmission, T, is defined as

$$T = \frac{I_2}{I_1} \tag{7.5}$$

where $I_2 < I_1$. The detection device computes a value of T thus giving a measure of the density of the image. The spot of light projected onto the film is moved in a raster scanning sequence as in television practice and the output sampled at given coordinate spacing.

This process will inevitably give rise to some distortion in the sampled image which may require correction before the final digitised image can be processed.

7.2.2 CODING

Where the processing of image data includes a transmission process it is essential to find an image coding system which can minimise the quantity of data for transmission with little apparent degradation of the image following reconstruction at the receiving end.

The digitised picture samples, $x_{i,k}$, are not stored or transmitted immediately but first coded into a sequence of binary words, $U_{i,k}$, of n bits where the number of grey levels is given by $N = 2^n$. The coding is arranged such that the volume of data for the coded sequence, d_c, is less than that for the input sequence, d_i. The efficiency of source coding is expressed by a **data compression factor**

$$K = \frac{d_i}{d_c} \tag{7.6}$$

Two methods of data compression are in general use, transform coding and threshold coding.

With transform coding the two-dimensional transform of an image is transmitted over the communication channel rather than the image itself. In many images there is a correlation of elements which causes the energy in the transformed image to be reduced to zero at near zero spatial frequencies. This acts to reduce the bandwidth necessary to achieve satisfactory transmission. The noise error immunity is good since each transformed element results from the weighted average of all the transformed samples. Thus a localised error is distributed over all the reconstructed elements and hence its subjective effect is reduced.

In one transform coding scheme the coding operates on subsets of the rows of data each block containing P samples of $x_{i,k}$ where P is the power of two. The blocks of samples are linearly transformed using a one-dimensional transformation into blocks of spectral coefficients, $u_{i,k}$. The number of samples transmitted is arranged to be less than the original samples in order to provide a K value in excess of unity.

Data compression may be achieved by taking advantage of the different variance for the sequence $u_{i,k}$ compared with the source data $x_{i,k}$. The variances of all the picture block elements are generally found to be of nearly equal magnitude so that each data element must be coded with the same number of bits. The variances of the transformed block elements on the other hand behave quite differently. Whilst some of these have quite large variance values the remaining ones become negligibly small. Thus we can either code each spectral element with a level of coding dependent on the variance value (variable block length coding) or eliminate those spectral coefficients belonging to the block elements which have small variance values below a given value (threshold coding).

Both methods find wide use in image coding and transmission systems.

7.2.3 IMAGE FUNCTIONS

In considering systems set up to process or generate two-dimensional information, it is necessary to define the performance of such a system in quantitative terms. This is equivalent to the requirement for systems' description for single-dimensional processing where a comparative set of terms is used.

For example the frequency response function of a two-dimensional system is defined as a **modulation transfer function**. This is given as the modulus of the Fourier transform of a two-dimensional matrix present at the output of the system and represents the effects of attenuation of a sinusoidal variation of the image content. Similarly, the **phase transfer function** is the phase of the Fourier transform of the output matrix and corresponds to the one-dimensional phase response of a system.

By far the most important of these concepts is that of the **point spread function** (PSF). This is equivalent to the impulse response of a two-dimensional system (Fig. 7.1) and would describe for example the distortion imposed by an optical system. If we let o(u,v) be a two-dimensional

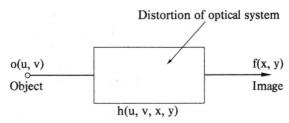

Fig. 7.1   The point spread function

function describing the object of interest then its image, f(x,y), can be stated as

$$f(x,y) = \int\int\limits_{-\infty}^{\infty} h(u,v,x,y)o(u,v)du,dv \qquad (7.7)$$

where (u,v) and (x,y) represent the spatial coordinates of the object and its image respectively.

   h(u,v,x,y) is the point spread function of the system producing the image f(x,y) (which could be for example a photographic plate). We can define the point spread function as the output produced by the system when its input or object is a point source of light, considered as a Dirac delta function. If the point spread function is invariant to shifts in the delta function, then

$$h(u,v,x,y) = h(x-u,y-v) \qquad (7.8)$$

Substitution in equation (7.7) gives the familiar form of a convolution equation showing that the image is really a convolution of the object with the point spread function describing the system. This has certain implications in image reconstruction which will be referred to later. Such systems are referred to as **linear isoplanar systems**.

7.2.4   COMPUTER IMAGE DOMAINS

In order to represent a digitised image in terms of the storage capability of a digital system we will adopt a formalised system defined in terms of a characteristic function known as an **image domain**. This has a value of 1 within an image and 0 outside of it. An image, or section of an image f(s), is a function f(i,j), defined for integer points (i,j), belonging to a set S(i,j), as the x,y coordinates; or alternatively line number and number of points along a line. The formal address of such an image stored within a computer system is the location, relative to some fixed origin in a computer store, of a list or set of lists which define S.

   The list could contain a pointer to another list giving grey level information for the coordinates. In a developed computer system for handling two-dimensional information, list manipulation systems would be used to enable space needed for the lists to be claimed and reclaimed dynamically.

7.3   Image Restoration and Enhancement

The concept of image restoration and enhancement forms a major two-dimensional signal processing activity.

Restoration may be defined as reconstruction of an image more closely approximating that of the object by inversion of some degradation phenomena. Two-dimensional data are rarely characteristic only of the object of interest. They will usually be found to be a linear superposition of the desired information with distortions and a noise factor. The restoration problem is how to separate out the required information from the superposition. The degradation of the image will imply distortions in the spatial frequency spectrum so that the restoration process will involve modification of the spectrum of the degraded image. This can be carried out by spatial filtering or by convolving the image with the point spread function of the image construction process. It is essential, therefore, that before any attempt at restoration is made some form of *a priori* knowledge concerning the degradation phenomena must be made available.

Enhancement is the attempt to improve the subjective appearance of an image or to give additional information about the object which was not apparent in the original image. Much improvement can be obtained by various forms of filtering designed to remove, for example, noise added to the original object during the process of image construction and digitisation or by emphasising the outline of some salient feature of the object image.

These methods may not be complex and can be simple to apply. One solution is low-pass filtering which would be obtained by averaging over a picture area through applying a running average first along the horizontal rows of a sampled and quantised image matrix and then along the vertical columns. This produces a stable filtered image which is related to the characteristics of the original image, i.e. retention of features greater than a given width expressed in samples.

More sophisticated methods are concerned with retaining or emphasising a particular feature of the data, such as edge enhancement, or by using colours to render the image of greater subjective value. Further discussion of some of these methods will be given later.

7.3.1   RESTORATION

The concept of computer image restoration is one of attempting to recover losses suffered in the process of producing the sampled and quantised image data. Many of these degradations are associated with the optical features of the process. Typical examples include camera motion, lens aberrations, low-pass characteristic of the electro-optical systems and atmospheric turbulence.

For linear shift invariant phenomena, including most of these degradations, the image can be restored by the use of inverse filtering. The restored output, g(x,y), is formed from the available image, f(x,y), and the point spread function of the degrading phenomenon, h(x,y), through the use of their Fourier transforms, viz.

$$G(m,n) = F(m,n) . H(m,n) \qquad (7.9)$$

where the transformed point spread function is defined as

$$H(m,n) = \frac{1}{2\pi} \int\limits_{-\infty}^{\infty}\int h(x,y)\exp(-j(mx+ny))dx,dy \qquad (7.10)$$

A further transformation gives the restored output in the spatial domain

$$G(m,n)g(x,y)$$

Experience has shown that image restoration by inverse filtering magnifies random errors because the higher frequencies are enhanced. This is because most optical systems tend to attenuate the higher spatial frequencies and, since h(x,y) will require to have reciprocal characteristics, the noise level of the system is increased. Thus the process must always be accompanied by noise suppression filtering [6] [7].

Where little is known of the point spread function a reasonable improvement in picture quality can be obtained by assuming a Gaussian transfer function which is acceptable for many disturbing phenomena. The compensating filter then becomes an inverse Gaussian curve which is implemented easily on the digital computer. The curve is circularly symmetric and defined by one parameter — the variance of the curve. This parameter can be adjusted until an acceptable contrast is obtained.

For complicated processes an **anisoplanar model** may be more relevant where the point spread function is different on each point on the two-dimensional plane. A convolution method is no longer applicable and it is necessary to evaluate equation (7.7) directly. Other models used for image enhancement include a multiplicative process in which the image is assumed to be comprised of the product of an illumination function i(x,y) and a reflected function r(x,y)

$$f(x,y) = i(x,y)r(x,y) \qquad (7.11)$$

A logarithmic operator is used to separate out the multiplication process followed by the linear filter.

A major difficulty in reproducing images acquired in the field is blurring of the image due to lens defects (defocusing) or camera motion. In most cases the cause of blurring can be modelled as a spatially invariant linear system. Thus we can model the blurred image $b(x,y)$ as the convolution of the object intensity function $f(x,y)$ and the point spread function $h(x,y)$ with additive noise $n(x,y)$ viz.

$$b(x,y) = f(x,y)*h(x,y) + n(x,y) \qquad (7.12)$$

Here $h(x,y)$ represents the PSF of the out-of-focus lens system or linear camera motion. The effects of these blurs can be seen if the Fourier transform equivalent of equation (7.12) is taken

$$B(u,v) = F(u,v).H(u,v) + N(u,v) \qquad (7.13)$$

where $H(u,v)$ is the two-dimensional frequency response of the blurring system and B, F and N are the transforms of the blurred image, original image and noise respectively.

By making certain assumptions regarding the nature of the blurring mechanism we can show that for a motion blur of length d whose direction is θ degrees off the horizontal, the frequency response has the form, sinc (df) where $f = u \cos \theta + v \sin \theta$ so that the higher frequencies are attenuated [6]. The frequency response for defocus blur takes the Bessel function form, $\frac{J_1(Rr)}{Rr}$, where R is the radius of the blur PSF and $r^2 = u^2 + v^2$. The shape is similar to motion blur response but is circularly symmetric. The problem resolves itself in the identification of the type of blur involved and then to deconvolve it from the blurred image to produce a restored image.

Identification has been attempted by a search for zero crossings of the frequency response of the blurring system [8] and through the use of cepstrum techniques [9]. The cepstrum was defined in Chapter 4 as the Fourier transform of the logarithm of $|B(u,v)|$. This implies that the convolution effects of $h(u,v)$ are additive in the cepstral domain and allows separation of the effects of blur.

However, the presence of noise mitigates the accurate identification of the characteristics of the two forms of blurring response. Modification of the cepstrum method by averaging through the power spectra has been proposed by Cannon to remove this difficulty [6].

Assuming the blurred signal and accompanying noise to be sample functions of a stationary random process then equation (7.13) can be expressed in power spectral terms as

$$S_b(u,v) = S_f(u,v)|H(u,v)|^2 + S_n(u,v) \qquad (7.14)$$

The method used by Cannon is to partition the blurred image $b(x,y)$ into smaller sub-images or pixels and to take the average over the square of the magnitude of the Fourier transforms of the sub-images. This constitutes an estimate of $S_b(u,v)$. Since $|H(u,v)|^2$ will be present in each one of the averaged sections with the contribution of the original image and noise varying from section to section a realistic estimation of $S_b(u,v)$ can be obtained.

A similar technique using the **power cepstrum** in place of the **power spectrum** leads to improved identification of motion and defocus blur when these are found together.

The **power cepstrum** is defined as

$$K_b(p,q) = F[\log_e S_b(u,v)] \tag{7.15}$$

where F indicates Fourier transformation. Neglecting noise equations (7.14) and (7.15) can be expressed by

$$K_b(p,q) = K_f(p,q) + K_h(p,q) \tag{7.16}$$

with $K_h(p,q)$ dominating the calculated result.

Some results of this technique applied to blurred images are reproduced in Figs. 7.2 to 7.7 contributed by Dr T. M. Cannon.* Figure 7.2 shows an original image blurred in the camera by an out-of-focus lens. The power cepstrum of this image is shown in Fig. 7.3 which shows the ring formation of spikes indicative of focus blur where the radius of the ring reveals the severity of the blur. Restoration of the image through deconvolution with the derived PSF is shown in Fig. 7.4. A similar image blurred through

Fig. 7.2   Deblurring techniques — original image blurred by an out-of-focus lens

*These results were produced at the University of Utah Computer Science Department, supported in part by contract no. DAHC15−73−C−0363 between the University and The Defense Advanced Research Projects Agency of the U.S. Department of Defense.

```
MAX  =      1  0
MIN  =      0  0
```

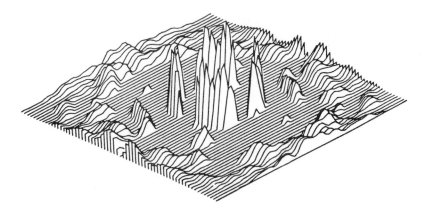

Fig. 7.3   Power cepstrum of Fig. 7.2

Fig. 7.4   Restoration of Fig. 7.2 through deconvolution

Fig. 7.5   Deblurring techniques — original image blurred by camera motion

MAX =   1 0
MIN =   0 0

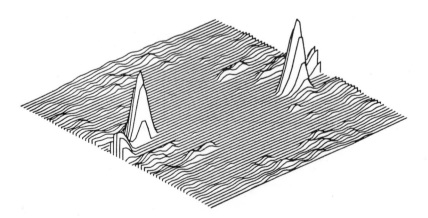

Fig. 7.6   Power cepstrum of Fig. 7.5

Fig. 7.7   Restoration of Fig. 7.5 through deconvolution

camera motion is shown in Fig. 7.5. The power cepstrum is shown in Fig. 7.6 where the twin peaks indicate motion blur and their separation and orientation denote severity and direction of the blur. Restoration of the image is shown in Fig. 7.7.

7.3.2   ENHANCEMENT

Improvement of a recorded image can be obtained in subjective terms by the use of computer enhancement techniques. Some of these techniques are a consequence of mathematical procedures, others are heuristic with emphasis on the subjective value of the improved image. In this section we will consider the following categories: intensity mapping; edge sharpening; frequency filtering; and pseudo-colour.

Intensity mapping is usually a non-linear operation

$$\hat{f}(x,y) = I(f(x,y)) \qquad (7.17)$$

Where $I(.)$ is a non-linear mapping of $f(.)$ independent of (x,y) in the picture content. Often the grey levels are not well distributed by the quantising process and two methods of non-linear mapping to correct this are found useful.

The first is to use a logarithmic device to stretch the quantisation over the grey scale to enhance a particular grey region [10]. The second is to use a histogram equalisation scheme in which the object is to produce a histogram as close to a uniform distribution as possible. In effect as many shades of grey as possible are used to display the image. Equalisation of this type is particularly valuable with X-ray images or images showing a histogram heavily biased towards one end of the grey scale [11].

Determining the boundary of objects in grey level pictures is a useful first step in extracting information from a recorded image. An edge is a region of the image where $f(x,y)$ has a gradient with a large magnitude. Various gradient operators have been devised and used for this purpose [12]. In many cases only the magnitude of the digital gradient has been used. For the purpose of edge tracking (contour tracking) directional information is also required [13]. The idea of tracing contours of a constant grey level in a given frame is useful since many of the points in one region will have the same grey level. If a contour is followed tracing these points then all that is necessary is for addressing information to be transmitted together with the grey level referred to. This reduces considerably the amount of data that need to be handled. The scheme is most useful when the number of grey levels is small or where an outline only of the image is necessary to convey the desired information.

A similar non-linear operation to intensity mapping is to modify the transformed frequency characteristic of the image. It has been found that a non-linear filter in the transform domain that tends to suppress large-value terms and consequently enhancing small-value terms results in a subjective improvement in image reconstruction. Because most of the energy of the image is found concentrated in the lower frequency portion of the Fourier domain or in the low sequency portion of the Walsh domain a logarithmic non-linear device reduces the smaller values less significantly than the larger values. These small-value coefficients occur in the higher frequency or higher sequency regions where most of the detail is to be found. A disadvantage of the method is that noise will also be enhanced and the acceptance of the method is dependent on the value of the enhanced detail that can be observed.

The objective with **pseudo-colour enhancement** is to increase the effective viewing dynamic range of the original grey scale by appealing to the human's visual response to colour. This is obtained by making a colour image from a monochrome image, mapping for example a particular spatial frequency range to a particular colour shade, or mapping a particular grey shade to a given intensity hue defining a colour shade. The amount of information that can be conveyed to the observer in this way can be quite considerable. It has been observed that although the human eye can differentiate only 20 or 30 different brightness levels in an image it is able to separate thousands of various colours [14]. A number of methods of implementation for these concepts are possible. One very simple technique which does not mix the primary colours involves mapping a separate portion of the grey scale into maximum brightness for each primary component. H. C. Andrews [15] has given a number of examples of this technique used for comparing X-ray images of a normal and a diseased lung with their black and white counterparts. A somewhat more complex pseudo-colour generation technique involves mixing of the primary colour components in accordance with a linear mapping. Here the pure primary colours occur only at the extreme ends and the exact centre of the grey scale. Another technique is to use a variable position window which colours the grey levels within that window all one colour leaving the rest of the image untouched. This would be useful for example in coloured contour mapping of the kind we find in conventional geographic maps.

7.4 Image Transformation

The essential feature of a transformed image is that a smaller number of significant values are realised than are found with the original spatial image. This underlines the value of transform coding for the transmission of digital images — fewer coded values are required to describe the transmitted image compared with the original. The use of a transformed image is valuable in other ways and, as we saw in the one-dimensional case, gives us a method of filtering the data economically on the digital computer through the use of a fast transform algorithm.

An image transformation can be viewed in several ways. It can be interpreted as a decomposition of the image data into a generalised two-dimensional spectrum. Each spectral component in the transform domain corresponds to the amount of energy of the spectral function within the original image. In this context the concept of frequency may now be generalised to include transformations of functions other than sine and cosine wave-forms, e.g. Walsh or Hadamard functions. This type of generalised spectral analysis is useful in the investigation of specific decompositions which are best suited for particular classes of images.

We can also consider an image transformation as a multi-dimensional rotation of coordinates. With the unitary transformations we shall be considering a major property is that measure is preserved. For example, the mean-square difference between two images is equal to the mean-square difference between the transforms of the image.

7.4.1 FOURIER TRANSFORMATION

We will consider first the Fourier transform and the fast Fourier transform algorithm applied to image processing. If we consider a digitised discrete spatial image $x_{i,k}$ where $i = 0,1,\ldots,N-1$, $k = 0,1,\ldots,M-1$ then the discrete Fourier transform evaluated at discrete angular frequency values

$$\omega_n = n\left(\frac{2\pi}{N}\right) \qquad \omega_m = m\left(\frac{2\pi}{M}\right) \qquad (7.18)$$

where
$$n, = 0,1,\ldots,N-1$$
$$m = 0,1,\ldots,M-1$$

is

$$X_{n,m} = \frac{1}{NM} \sum_{i=0}^{N-1} \sum_{k=0}^{M-1} x_{i,k} \exp[-j(\omega_n i + \omega_m k)] \qquad (7.19)$$

and the inverse transform is given by

$$x_{i,k} = \sum_{n=0}^{N-1} \sum_{m=0}^{M-1} X_{n,m} \exp[j(\omega_n i + \omega_m k)] \qquad (7.20)$$

Even though $x_{i,k}$ is a real positive function, its transform $X_{n,m}$ is in general complex, thus while the image contains NM components the transform contains 2NM components, the real and imaginary or magnitude and phase components of each spatial frequency. However, since $x_{i,k}$ is a real positive function, $X_{n,m}$ exhibits a property of conjugate symmetry, viz.

$$X_{n,m} = -X_{-n,-m} \qquad (7.21)$$

Consequently

$$X_{n,m} = X^*_{-n,-m} \qquad (7.22)$$

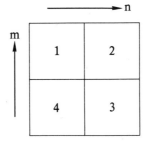

Fig. 7.8   Conjugate symmetry of a two-dimensional Fourier transform

Figure 7.8 illustrates the conjugate symmetry property of the Fourier transform when the zero spatial frequency term is located at the centre of the transform plane. This is accomplished by multiplying the image by function $(-1)^{i+k}$ before the transformation. Samples in quadrants (1) and (3) are complex conjugates of one another as are samples in quadrants (2) and (4). Due to the conjugate symmetry property of the Fourier transform it is only necessary to transmit the samples of one-half of the transform plane; the other half can be reconstructed from the half plane sample transmitted.

Hence, the Fourier transform of an image can be described by NM data components.

As in the one-dimensional case for a Fourier series representation to be valid the field must be periodic. We must therefore consider the regional image periodically repeated in the horizontal and vertical directions as shown in Fig. 7.9.

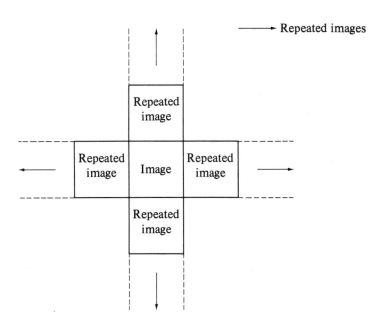

Fig. 7.9   Effective periodic repetition of a digitised field

While the Fourier transform possesses many desirable analytic properties, it has two major drawbacks: complex rather than real number computations are necessary and the rate of convergence is low. This latter disadvantage is significant in image coding. Other transforms which do not have these major drawbacks are the cosine and sine transforms and the non-sinusoidal Hadamard or Walsh transforms.

7.4.2   THE COSINE TRANSFORM*

As shown earlier the Fourier series representation of any continuous real and symmetric function contains only real coefficients corresponding to the cosine terms of the series. In image processing this condition can be realised by adjusting the image field such that it is folded about its edges symmetrical about a point, $i = -1/2$ and $k = -1/2$ (Fig. 7.10) [16] using the relationship

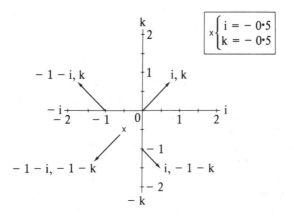

Fig. 7.10   The cosine transform field

$$x_s(i,k) = \begin{array}{lll} x(i,k) & i \geqslant 0 & k \geqslant 0 \\ x(-1-i,k) & i < 0 & k \geqslant 0 \\ x(i,-1-k) & i \geqslant 0 & k < 0 \\ x(-1-i,-1-k) & i < 0 & k < 0 \end{array} \qquad (7.23)$$

Taking a Fourier transform about this point of symmetry gives

$$X_s(m,n) = \frac{1}{2N} \sum_{i=-N}^{N-1} \sum_{k=-N}^{N-1} x_s(i,k)\exp\left\{ -j\frac{\pi}{N}(m(i+{}^1\!/_2)+n(k+{}^1\!/_2)) \right\}$$

$$(7.24)$$

for $m,n = -N \ldots -1,0,1 \ldots N-1$

Since m goes from $-N$ to $+N$ and consequently brings about a sign change in $\sin\frac{\pi}{N}(m(i+{}^1\!/_2))$ and $\cos\frac{\pi}{N}(m(i+{}^1\!/_2))$ equation (7.24) reduces to

$$X_s(m,n) = \frac{2}{N} \sum_{i=0}^{N-1} \sum_{k=0}^{N-1} x(i,k)\cos\left[\frac{\pi}{N}m(i+{}^1\!/_2)\right]\cos\left[\frac{\pi}{N}n(k+{}^1\!/_2)\right]$$

$$(7.25)$$

which is the even **cosine transform** for the normalised version of equation (7.24) when $n,m \neq 0$.

When $n,m = 0$ then

$$X_s(m,n) = \frac{1}{N} \sum_{i=0}^{N-1} \sum_{k=0}^{N-1} x(i,k) \qquad (7.26)$$

It can be shown that equation (7.25) can be expressed as

$$X_s(m,n) = \frac{2}{N} Re \left\{ exp[j \frac{\pi m}{2N}] \sum_{i=0}^{N-1} \sum_{k=0}^{N-1} x(i,k)exp[-j \frac{\pi}{N}(mi+nk)] \right\}$$

$$(7.27)$$

Where Re(.) implies the real part of the term enclosed.

From (7.27) it follows that all the N^2 coefficients can be computed using a two-dimensional fast Fourier transformation over 2N points. Similarly a discrete sine transform can be defined by replacing the Re(.) of equation (7.27) by Im(.) which denotes the imaginary part of the term enclosed. It has been shown that the performance of the discrete cosine transform in image processing compares closely with that of the **Karhunen–Loeve transform**, known to approach the optimal in terms of variance, distribution, mean-square error criteria and rate distortion factor but which cannot be calculated rapidly by means of a fast algorithm [17].

7.4.3   WALSH TRANSFORMATION*

For symmetric Walsh matrices of order $N = 2^P$, the two-dimensional Walsh transform may be written in series form as

$$X_{n,m} = \frac{1}{N} \sum_{i=0}^{N-1} \sum_{k=0}^{N-1} x_{i,k}(-1)^{P(i,k,m,n)} \qquad (7.28)$$

Where

$$P(i,k,m,n) = \sum_{b=0}^{P-1} (m_b i_b + n_b k_b) \qquad (7.29)$$

Here the terms m_b, i_b, n_b, and k_b are the bit states of the binary representations of m, i, n and k respectively. Evaluation of equation (7.28) is obtained in the same way as that described for the Fourier transform, namely decomposition using a partial transformation

$$X_{m,k} = \sum_{i=0}^{N-1} x_{i,k}WAL(n,i) \qquad (7.30)$$

This is followed by a second single-dimensional transformation

$$X_{m,n} = \sum_{k=0}^{N-1} X_{m,k}WAL(m,k) \qquad (7.31)$$

Where WAL(n,i) and WAL(m,k) represent the one-dimensional Walsh transformation series corresponding to the bit operation given by equation

(7.29). The zero sequency term for equation (7.28) is a measure of the average value for the summation of terms in the data matrix. Thus we can write

$$X_{0,0} = \sum_{i=0}^{N-1} \sum_{k=0}^{N-1} x_{i,k} \tag{7.32}$$

If $x_{i,k}$ represents a positive real function then the maximum possible value for $x_{0,0}$ is $N^2 A$ where A is the maximum value of the function. All the Walsh domain samples other than $x_{0,0}$ will vary between $\pm N^2 . A/2$ thus establishing a bound for all other Walsh domain samples.

As with the single-dimensional transform, Parseval's relationship can be applied giving

$$\sum_{i=0}^{N-1} \sum_{k=0}^{N-1} |x_{i,k}|^2 = \frac{1}{N^2} \sum_{m=0}^{N-1} \sum_{n=0}^{N-1} |X_{m,n}|^2 \tag{7.33}$$

This has important implications in bandwidth reduction, since, if a few of the Walsh domain samples are of large magnitude then it follows that the remainder will be of small magnitude. The small magnitude samples may be discarded to achieve a reduced bandwidth for the data sample.

7.4.4 COMPUTATION OF THE DISCRETE TWO-DIMENSIONAL FOURIER TRANSFORM

Equation (7.28) can be re-written in a form suitable for computation as

$$X_{n,m} = \sum_{i=0}^{N-1} \sum_{k-0}^{N-1} x_{i,k} W_1^{-ni} W_2^{-mk} \tag{7.34}$$

Where $x_{i,k} = x(nT,mS)TS$ \hfill (7.35)

and T and S are the sampling periods in x and y direction respectively

$$W_1 = \exp(j \frac{2\pi}{N})$$

$$\tag{7.36}$$

$$W_2 = \exp(j \frac{2\pi}{M})$$

Similarly the inverse transformation equation (7.20) can be expressed as

$$x_{i,j} = \frac{1}{MN} \sum_{n=0}^{N-1} \sum_{m=0}^{M-1} X_{n,m} W_i^{-ni} W_2^{-mk} \tag{7.37}$$

Equation (7.34) can be evaluated through a one-dimensional transformation by partitioning into the following form

$$X_{n,m} = \sum_{i=0}^{M-1} \left[\sum_{k=0}^{N-1} x_{i,k} W_1^{-ni} \right] W_2^{-mk} \tag{7.38}$$

The inner summation is calculated first, then the outer summation. This means that the calculation of the two-dimensional transform can proceed in two steps. First we calculate the one-dimensional Fourier transform of each row of the data $x_{i,k}$. Then, we calculate the one-dimensional Fourier transform of each column of the resulting array from the first step. Using a fast Fourier transform algorithm to do the one-dimensional transforms, then the total number of complex multiplications and additions required for the complete two-dimensional transform is $2(MN)\log_2(MN)$.

7.5 Two-Dimensional Digital Filtering

Two-dimensional digital filtering finds applications in many areas of image processing including enhancement, restoration and feature extraction. As shown in Chapter 6 digital filtering represents a general linear filtering operation performed on sampled data.

Two-dimensional filters can be defined by a similar relationship

$$y_{(i,k)} = \sum_{n=0}^{N-1} \sum_{m=0}^{M-1} a(n,m)x(i-n,k-m) - \sum_{p=0}^{P-1} \sum_{q=0}^{Q-1} b(p,q)y(i-p,k-q)$$

$$\tag{7.39}$$

where $x(i,k)$ represents the input data, $y(i,k)$ the output data and $a(n,m)$ and $b(p,q)$ are the coefficients defining the digital filter. As in the one-dimensional case if $a(n,m)$ and $b(p,q)$ are different from zero, the digital filter has infinite impulse response. If all $b(p,q)$ are zero then the filter has finite response.

FIR digital filters can be designed with an exact linear phase and always permit a stable implementation. In general the design methods represent extensions from the one-dimensional theory with due regard to the larger size of data matrix handled in the two-dimensional case. Fusco [18] has developed ways of convolving large matrices equivalent to the one-dimensional methods of Stockham described earlier (Chapter 6). Other design methods include work on frequency sampling and optimum filters by Hu and Rabiner [19], window technique by Huang [20] and conversion algorithms from the one-dimensional case described by McClellan [21].

The choice between the use of IIR or FIR designs depends very much on the application. Many applications will best be served by the non-recursive design where the filters will always be stable and can be designed for zero or linear phase characteristic which is important for image processing. Furthermore one-dimensional design methods can be extended to the two-dimensional case and filters of this kind can be designed quite quickly.

Two-dimensional IIR digital filter design cannot be transformed easily from the one-dimensional case due to a fundamental theorem of algebra which does not extend to the two-dimensional case. Among the various methods of design are those involving transformation from the Laplace s-plane to the z-plane, mapping from the continuous domain, and transformation from the one-dimensional case. In all of these the testing and correcting for stability plays an important role.

Whilst the operation of IIR two-dimensional filters requires fewer arithmetic operations the design work necessary to achieve a stable filter can be extensive. They are particularly useful in such applications as seismic data processing where the volume of data is large or real-time conditions prevail and the extra computing time needed in the design phase is no disadvantage.

7.5.1 TWO-DIMENSIONAL CONVOLUTION

The convolution properties deduced from the discrete Fourier transform of a one-dimensional sequence are also applicable to the two-dimensional case. Thus for a one-dimensional series having N values, x_i convolved with a function (second one-dimensional series of M values) y_k then the discrete circular convolution would be

$$C_i = \sum_{k=0}^{N-1} y_k x_{(i-k)} \qquad (7.40)$$

Where
$$i = 0,1,\ldots,N-1$$
$$k = 0,1,\ldots,M-1$$

Similarly for a two-dimensional series, $x_{i,k}$ consisting of an NxM matrix convolved with a similar matrix series $y_{u,v}$ we have

$$C_{i,k} = \sum_{u=0}^{N-1} \sum_{v=0}^{N-1} y_{u,v} x_{(i-u)(k-v)} \qquad (7.41)$$

where;
$$i,u = 0,1,\ldots,N-1$$
$$k,v = 0,1,\ldots,M-1$$

It can be shown also that the convolution sum may be evaluated rapidly using a two-dimensional fast Fourier transform. The procedure is to obtain the fast Fourier transform of each matrix, viz.

$$X_{m,n} = \sum_{i=0}^{N-1} \sum_{k=0}^{N-1} x_{i,k} W_N^{-im} W_N^{-kn} \tag{7.42}$$

where $W_N = \exp\left(j\,\dfrac{2\pi}{N}\right)$ $\tag{7.43}$

and similarly for $y_{u,v}$ giving $Y_{p,q}$.

The product $X_{m,n} \cdot Y_{p,q}$ is computed and the inverse transform evaluated. This is equivalent to the convolution given by equation (7.41)

$$C_{i,k} = \frac{1}{N^2} \sum_{r=0}^{N-1} \sum_{s=0}^{N-1} D_{r,s} W_N^{ir} W_N^{ks} \tag{7.44}$$

where $[D_{r,s}] = [X_{m,n}] \cdot [Y_{p,q}]$

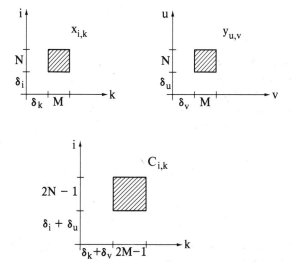

Fig. 7.11   Displacement of a two-dimensional convolved matrix

The series $x_{i,k}$ and $y_{u,v}$ may not commence from zero in a finite two-dimensional interval. Figure 7.11 illustrates the convolution operation where the series are displaced by $\delta_i = \delta_u$ and $\delta_k = \delta_v$ respectively. This gives a convoluted matrix displaced relative to the original series.

As with the one-dimensional case, to avoid distortion due to circular convolution, it is necessary to take only one period of the result and to translate it in the i dimension by $\delta_i + \delta_u$ and in k by $\delta_k + \delta_v$.

There are a number of filtering operations that do not require the convolution property. One example is where the data matrix consists of additive signal and noise components. An optimum filter can be designed through a two-dimensional transformation of a matrix based on the known covariance matrices of the signal and noise to give a minimum mean-square error for the filtered signal (see Chapter 6). Optimum filter design of this kind results in filter weights which have levels other than zero or one. This presents difficulties when using non-sinusoidal transforms, e.g. Walsh or Haar, and sub-optimum methods are employed where those filter coefficients having small values close to zero are replaced by zero values.

An example of this is given in Fig. 7.12. Here the chequerboard pattern shown in (*a*) has added to it 5 db of random noise with the result shown in (*b*). The noisy data are transformed and those coefficients having a modulus value less than 2 per cent of the peak value are replaced by zeros. The modified matrix is inversely transformed to give the improvement shown in (*c*). This comparatively simple technique of threshold filtering using the Walsh transform has wide applications in image processing.

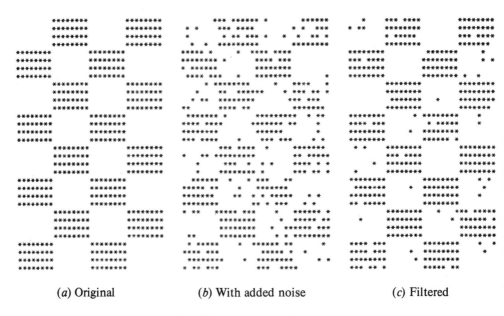

(*a*) Original   (*b*) With added noise   (*c*) Filtered

Fig. 7.12 Threshold filtering

7.5.2 FIR FILTERING

FIR digital filters are inherently stable since their z-transform is a finite polynomial (see Chapter 6). Thus, design techniques in one-dimension are often directly extensible into the two-dimensional region. Further, it is possible to convert an actual one-dimensional filter design into a two-dimensional realisation by means of a transformation algorithm. In this section we will commence with a general introduction to the theory of two-dimensional FIR filters and then consider an extension of some of the design techniques described in Chapter 6 into two-dimensions including the transformation method. The two-dimensional z-transform of the filter is defined as

$$H_{(z_1,z_2)} = \sum_{n=0}^{N-1} \sum_{m=0}^{M-1} h(n,m)z_1^{-n}.z_2^{-m} \tag{7.45}$$

where $h(n,m)$ is the impulse response (point spread function) of the filter and n,m are finite sequences. This is related to the spatial frequency response through the transformation, $z_1 = \exp(j\omega_1)$ and $z_2 = \exp(j\omega_2)$, viz.

$$H_{(w_1,w_2)} = \sum_{n=0}^{N-1} \sum_{m=0}^{M-1} h(n,m)\exp[-j(\omega_1 n + \omega_2 m)] \tag{7.46}$$

The two-dimensional FIR filter may be realised in several ways. The simplest technique is its realisation through direct convolution, i.e.,

$$y(i,k) = \sum_{n=0}^{N-1} \sum_{m=0}^{M-1} h(n,m)x(i-n)\,(k-m) \tag{7.47}$$

where $x(i,k)$ is the input of the filter and $y(i,k)$ is the filter output. This result can be obtained indirectly via the fast Fourier transform as described in Section 7.5.3.

A major problem in the processing of images using the convolution method is the large information content (typically 1000×1000 discrete points). It is not normally possible for the entire image to be processed simultaneously within the computer memory and methods have been developed by which the convolution of a filter with a very large signal can be carried out by successive convolutions of the filter with part of the signal. This process was described in some detail earlier in Section 6.2.1.

An 'overlap-save' method applicable to the two-dimensional case is due to Fusco [18] and is illustrated in Fig. 7.13. Here an image $x_{i,k}$ is considered shown fivided into 25 sub-areas. It is non-zero in the two-dimensional area $M = M_i \times M_k)$ where M is much greater than the area $N = (N_i \times N_k)$ in which the values $h_{i,k}$ defining the filters are given. In order to realise the fast convolution of $x_{i,k}$ with $h_{i,k}$ using the overlap-save technique, the following steps need to be carried out:

1. The digitised image is subdivided into an appropriate number of two-

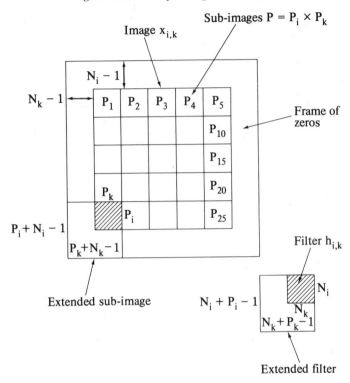

Fig. 7.13 The 'overlap-save' method of FIR filtering

dimensional sub-images of area $P = (P_i \times P_k)$ elements with P containing an arbitrary number of pixels.

2. The original sequence is extended with a frame of zeros of dimension at least $(N_i - 1)$ and $(N_k - 1)$.

3. Each sub-image of P pixels, extended with a frame of zeros of dimension at least $(N_i - 1)$ and $(N_k - 1)$ is convolved with the filter which has also been extended with a frame of zeros $(P_i - 1)$ and $(P_k - 1)$ corresponding to the sub-image.

4. Only the points found within the location of the sub-area $(P_i \times P_k)$ within the resulting image are retained as useful results of the convolution.

5. Steps **(3)** and **(4)** are repeated until the complete image is convolved.

The problem in designing two-dimensional FIR filters can be considered as a linear approximation problem and linear programming methods are applicable. However, linear programming has not been applied widely to the design of such filters due to its low efficiency when optimising over the large number of parameters required. For example, in the case of a matrix filter of 31 × 31 values then optimising over 16 × 16 = 256 parameters would be needed and linear approximation in such a case is extremely slow. For this reason much attention has been given to sub-optimum filter design methods which are computationally more efficient although not providing such good solutions to the approximation problem. Amongst these is the frequency sampling technique which is described below.

This depends on the initial specification of filter frequency response at uniformly spaced frequencies. The discrete Fourier transform relations for the filter may be obtained by evaluating equation (7.46) at the discrete set of values

$$\omega_{k_1} = k_1 \left(\frac{2\pi}{N}\right) \quad k_1 = 0,1,\ldots,N-1$$

$$\omega_{k_2} = k_2 \left(\frac{2\pi}{M}\right) \quad k_2 = 0,1,\ldots,M-1 \tag{7.48}$$

giving

$$H(\omega_{k_1},\omega_{k_2}) = \sum_{n=0}^{N-1} \sum_{m=0}^{M-1} h(n,m)\exp[-j(n\omega_{k_1}+m\omega_{k_2})] \tag{7.49}$$

The inverse Fourier transform is readily obtained as

$$h(n,m) = \frac{1}{NM} \sum_{k_1=0}^{N-1} \sum_{k_2=0}^{N-1} H(\omega_{k_1},\omega_{k_2})\exp[j(n\omega_{k_1}+m\omega_{k_2})] \tag{7.50}$$

A frequency interpolation formula from the discrete Fourier transform coefficients may be derived by inserting equation (7.50) into (7.46) giving

$$H(\exp(j\omega_1)\exp(j\omega_2)) = \sum_{n=0}^{N-1} \sum_{m=0}^{M-1} \left[\frac{1}{NM} \sum_{k_1=0}^{N-1} \sum_{k_2=0}^{M-1} H(\omega_{k_1},\omega_{k_2})\exp[j(n\omega_{k_1}+\right.$$

$$\left. m\omega_{k_2})]\exp[-j(n\omega_1+m\omega_2)] \right] \tag{7.51}$$

Interchanging orders of summation and summing over the n and m indices gives

$$H(\exp(j\omega_1)_1\exp(j\omega_2)) = \sum_{k_1=0}^{N-1} \sum_{k_2=0}^{M-1} H(\omega_{k_1},\omega_{k_2})A(k_1,k_2,\omega_1,\omega_2) \tag{7.52}$$

where

$$A(k_i,k_2,\omega_1,\omega_2) = \frac{1}{NM}\left[\frac{1-\exp(-jN\omega_1)}{1-\exp[j(\omega_{k_1}-\omega_1)]}\right] \cdot \left[\frac{1-\exp(-jM\omega_2)}{1-\exp[j(\omega_{k_2}-\omega_2)]}\right]$$

(7.53)

Equation (7.53) can be used as a basis for designing frequency sampling two-dimensional filters. We see from equation (7.53) that the continuous frequency response of the filter is a linear combination of shifted interpolating functions $A(k_i, k_2, \omega_1, \omega_2)$ weighted by the discrete Fourier transform coefficients $(H(k_1, k_2)$. These latter coefficients are called frequency samples which exactly specify the value of the frequency response required from the filter at uniformly spaced frequencies. In the design of frequency sampling filters the majority of frequency samples are given specific values dependent on the frequency response being approximated. The remaining unspecified frequency samples are left as free variables to be optimised according to some minimisation criteria. Details of this method are discussed by Hu and Rabiner [19].

An important method of two-dimensional filtering design involves transforming a one-dimensional filter into a two-dimensional filter by a change of variables. The method is applicable to a limited but important class of filters including low-pass and band-pass circularly symmetric filters. The advantages obtained are speed of computation and ability to use existing knowledge of one-dimensional filter design and performance. In some cases the design can be shown to be optimal unlike other two-dimensional methods discussed above.

The frequency response of a discrete two-dimensional digital filter was defined in equation (7.49). Following McClellan [21] if we now constrain the impulse response to $N \times M$ where N and M are odd integers and impose the symmetry conditions

$$h(N-p-1,m) = h(p,m) \quad p = 0,1,\ldots,N_1 = \tfrac{1}{2}(N-1)$$
$$h(n,M-p-1) = h(n,p) \quad p = 0,1,\ldots,M_1 = \tfrac{1}{2}(M-1)$$

(7.54)

then the frequency response given by equation (7.49) can be expressed as

$$H(\omega_{k_1},\omega_{k_2}) = \exp[-j(\omega_{k_1}N_1+\omega_{k_2}M_1)] \sum_{n=0}^{N_1} \sum_{m=0}^{M_1} a(n,m)\cos \omega_{k_1}n \cos \omega_{k_2}m$$

(7.55)

where $a(n,m)$ expresses the impulse response matrix

$$\begin{aligned}
a(0,0) &= h(N_1M_1) \\
a(0,m) &= 2h(N_1,M_1-m) & m &= 1,\ldots,M_1 \\
a(n,0) &= 2h(N_1-n,M_1) & n &= 1,\ldots,N_1 \\
a(n,m) &= 4h(N_1-n,M_1-m)
\end{aligned}$$

(7.56)

This reduces to

$$\hat{H}(\omega_{k_1},\omega_{k_2}) = \sum_{n=0}^{N_1} \sum_{m=0}^{M_1} a(n,m)\cos \omega_{k_1}n \cos \omega_{k_2}m \qquad (7.57)$$

Since $\exp[j(\omega_{k_1}N_1+\omega_{k_2}M_1)]$ has unit magnitude.

A similar simplification can take place with a one-dimensional filter having an impulse response $h(n)$, $= 0,1,\ldots,N-1$. The frequency response is defined by the Fourier transform

$$G(\omega_k) = \sum_{n=0}^{N-1} h(n)\exp(-j\omega_k n) \qquad (7.58)$$

Where the length of the impulse response is odd, $N = 2N_1 + 1$, and symmetric, $h(n) = h(N-1-n)$, $n = 0,1,\ldots,N_1$ this reduces to

$$G(\omega_k) = \exp(-j\omega_k N_1) - \sum_{n=0}^{N_1} b(n)\cos \omega_k n \qquad (7.59)$$

Where $b(0) = h(N_1)$ and $b(n) = 2h(N_1-n)$, $n = 1,\ldots,N_1$
Under these conditions

$$\hat{G}(\omega_k) = \sum_{n=0}^{N_1} b(n) \cos.\omega_k n \qquad (7.60)$$

We can also write this equation as

$$\hat{G}(\omega_k) = \sum_{n=0}^{N_1} \hat{b}(n) (\cos \omega_k)^n \qquad (7.61)$$

By the choice of suitable coefficients for $\hat{b}(n)$.

The change of variables suggested by McClellan depends on the fact that $\hat{G}(\omega_k)$ and $\hat{H}(\omega_{k_1},\omega_{k_2})$ are both sums of cosine functions. The following change of variables is proposed

$$\cos \omega_k = A \cos \omega_{k_1} + B \cos \omega_{k_2}$$
$$+ C \cos \omega_{k_1}.\cos \omega_{k_2} + D \qquad (7.62)$$

Substitution of equation (7.62) into equation (7.61) yields a bi-variate polynomial in $\cos \omega_{k_1}$ and $\cos \omega_{k_2}$ viz.

$$\sum_{n=0}^{N_1} \sum_{m=0}^{N_1} \hat{a}(n,m) (\cos \omega_{k_1})^n(\cos \omega_{k_2})^m \qquad (7.63)$$

which can be written uniquely in the same form as $H(\omega_{k_1}, \omega_{k_2})$ namely

$$\sum_{n=0}^{N_1} \sum_{m=0}^{N_1} a(n,m) \cos \omega_{k_1} n \cos \omega_{k_2} m \qquad (7.64)$$

Thus the proposed change of variables will preserve the proper form for a two-dimensional filter taking the form of a square array (i.e., $N = M$).

The change of variables given by equation (7.62) defines a mapping from the interval $(0,0\cdot5)$ of the one-dimensional frequency axis to the square $(0,0\cdot5) \times (0,0\cdot5)$ in the two-dimensional frequency plane. It is necessary only to determine the values of the constants A,B,C and D given in this equation.

Solving equation (7.62) for ω_{k_2} as a function of ω_{k_1} gives

$$\omega_{k_2} = \cos^{-1} \left[\frac{\cos \omega_k - D - A \cos \omega_{k_1}}{B + C \cos \omega_{k_1}} \right] \qquad (7.65)$$

As shown in Fig. 7.14 we see that for a fixed angular frequency we get a curve in the $(\omega_{k_1}, \omega_{k_2})$ plane and along this curve the transformed two-dimensional frequency response is a constant equal to the value of the one-dimensional frequency response at ω_k. This diagram is obtained by putting $A = B = C = -D = 0\cdot5$ which gives a family of contours which completely describe the transformed two-dimensional frequency response and may be used to design circularly symmetric filters.

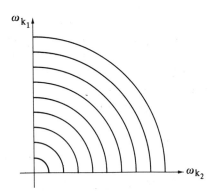

Fig. 7.14   Contour lines for a circularly symmetric filter

In order to ensure that the mapping from ω_k to $(\omega_{k_1}, \omega_{k_2})$ is well defined it is necessary that A,B,C, and D be chosen such that the right-hand side of equation (7.62) lies between $+1$ and -1 for all angular frequencies between 0 and $0\cdot5$. Thus $\omega_k = 0$ is mapped to $(0,0)$ and $\omega_k = 0\cdot5$ to $(0\cdot5, 0\cdot5)$, or vice versa, thus

$$\pm 1 = A + B + C + D$$

$$\mp 1 = -A - B + C + D \qquad (7.66)$$

which simplifies to A = ± 1 − B and C = − D reducing the number of parameters required to choose from the four to two. The best choice for A and C to approximate circular symmetry is 0·5.

An example of two-dimensional FIR filtering enhancement to a nuclear scintigraph is shown in Fig. 7.15. The original digitised image, shown in Fig. 7.15(a) was obtained from X-ray scanning of the liver region and consists of a 64 × 64 matrix of pixels having an 8 bit quantised grey level. The method of acquisition involves considerable frequency distortion of the desired information which the filtering process is designed to remove. The data are subject to a two-dimensional band-pass filter defined by equation (7.45) where h(n,m) are the samples of an impulse-response function obtained by Fourier transformation of the desired rectangular band-pass frequency response. In deriving this the ideal frequency response function is first modified by a window function W(n,m). This is obtained from the circular rotation of a one-dimensional window of the Lanczos type [22] which minimises the ripple outside the pass band. Cappellini [23] has applied this technique to the original data using carefully selected filter parameters. The result is shown in Fig. 7.15(b) and indicates the presence of a dense growth or cyst which is not apparent in the original distorted data.

7.5.3 IIR FILTERING

The problems associated with two-dimensional filtering are considerably more formidable than the 1−D case. The difficulties relate to the absence of a fundamental algebraic theorem for polynomials in two independent variables. A definition was given by equation (7.39) where a(m,n) and b(p,q) both have finite values. The transfer function may be written in z-transform notation as

$$H_{(z_1,z_2)} = \frac{Y_{(z_1,z_2)}}{X_{z_1,z_2}} = \frac{\sum_{n=0}^{N-1}\sum_{m=0}^{M-1} a_{(n,m)} z_1^{-n} z_2^{-m}}{\sum_{p=0}^{P-1}\sum_{q=0}^{Q-1} b(p,q) z_1^{-p} z_2^{-q}} \qquad (7.67)$$

The design problem is to choose coefficients for a(n,m) and b(p,q) which will allow a stable recursive realisation.

If the filter is unstable, any noise (including round-off errors) can propagate through to the output and be amplified. Furthermore, in inverse

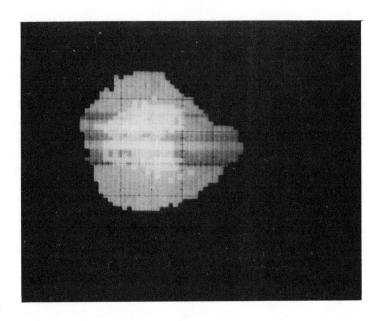

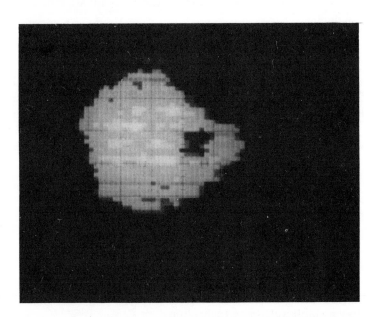

Fig. 7.15　Example of 2-D filtering applied to a nuclear scintigraphy of a liver
(*a*) original image, (*b*) processed image

filtering for image restoration, the initial conditions are generally unknown. Consequently it is desirable that the recursive filtering operation be relatively independent of the initial conditions. Therefore, in addition to stability we may require a rapidly decaying impulse response.

One-dimensional recursive filters can be realised efficiently by factorising their z-transforms into a cascade of real or complex pole-zero networks and converting these simple sections into difference equations acted on in sequence. A test for stability can easily be made by finding the poles of the z-transform function and the filter stabilised by replacing the poles in the instability region with poles in conjugate reciprocal positions with respect to the unit circle. For the two-dimensional case it is not possible to factorise the denominator of equation (7.67) so that these design and stability methods no longer apply. A complex polynomial has to be evaluated having many coefficients which need to be truncated for computer evaluation to a reasonable number of binary bits. This can lead to cumulative errors and instability even if the criterion for bounded input-output stability is realised, namely

$$B_{(z_1,z_2)} = \sum_{p=0}^{P-1} \sum_{q=0}^{Q-1} b_{(p,q)} z_1^{-p} z_2^{-q} \neq 0 \qquad (7.68)$$

A basic theorem for satisfying this criterion is due to Shanks [24].

Given that $B(z_1,z_2)$ is a polynomial in (z_1,z_2), for the coefficients of the expansion of $1/B(z_1,z_2)$ in positive powers of z_1 and z_2 to converge absolutely, it is necessary and sufficient that $B(z_1,z_2)$ not be zero for $|z_1|$ and $|z_2|$ simultaneously less than or equal to 1. This is analogous to the one-dimensional case for stability which requires that no poles are located outside the unit circle.

A simpler stability test is due to Huang [25]. He states that a filter

$$H(\omega,z) = \frac{A(\omega,z)}{B(\omega,z)} \qquad (7.69)$$

is stable if, and only if, the image of the unit circle $|z| = 1$ mapped into the ω plane by $B(\omega,z) = 0$ does not intersect with the region $|\omega| \geq 1$ and in addition no point in the region $|\omega| \geq 1$ maps into the point $z = \infty$. In order to apply this test the z unit circle must be mapped into the ω plane by $B(\omega,z) = 0$. For the filter to be stable, its image must not intersect at a point $|\omega| \leq 1$. Further a solution must be found for $B(\omega,z) = 0$ for ω, setting $z^{-1} = 0$. The problem with both of these tests is that, in theory, an infinite amount of computational time is required to satisfy them. Attempts have been made to find an implementation requiring only a finite amount of time by Huang [25], Anderson and Jury [26] and Maria and Fahmy [27]. The Anderson—

Jury algorithm appears to be the most efficient but still represents a considerable computational task.

The purpose of using these tests is of course to implement them into the design methods for a two-dimensional recursive filter. An alternative method, which functions extremely well in the one-dimensional case, is to attempt to stabilise an unstable filter. This can be achieved in the one-dimensional case since the poles of the system located outside the unit circle (if any) can be replaced by poles in conjugate reciprocal locations within the unit circle. This cannot be done in two dimensions simultaneously, consequently stabilisation is a fairly complex process. Read and Treitel [28] have defined a two-dimensional discrete Hilbert transform technique to provide this stabilisation. Similar attempts using the notation of a two-dimensional complex cepstrum have been made by Pistor [29] and Dudgeon [30]. The two-dimensional cepstrum is similar to the one-dimensional definition, namely the inverse Fourier transform of the complex logarithm of the Fourier transform of the sequence (see Chapter 4). In general terms the cepstrum is used to decompose the magnitude-squared frequency response of a potentially unstable filter to derive a stable recursive two-dimensional filter.

The difficulty with these methods is that exact decomposition yields a denominator polynomial which is not of finite degree although the denominator of the original magnitude-squared function is itself finite.

In several specific cases two-dimensional arrays can be represented exactly as one-dimensional sequences. Mersereau and Dudgeon [31] demonstrate the conditions under which this is possible and hence evaluate the coefficients for the two-dimensional numerator and denominator polynomial from well-known one-dimensional design techniques using a given impulse-response function. Other translated techniques from the one-dimensional case are possible. The designer of two-dimensional IIR filters faces two fundamental problems: their stability and their synthesis. Three approaches are currently being adopted:

1. To incorporate a stability test into the design procedure.
2. To approximate a suitable filter function and then to carry out a stabilisation procedure.
3. To transform a stable one-dimensional filter into a two-dimensional form using some form of angular transformation.

Huang's simplification of the stability problem may be stated in terms of $B(z_1, z_1) \neq 0$ for $| z_1 | < 1$ or $| z_2 | < 1$ if and only if the following two conditions hold

$$B(z_1, 0) \neq 0 \qquad | z_1 | \leq 1 \qquad\qquad (7.70)$$

$$B(z_1, z_2) \neq 0 \qquad | z_1 | = 1, \qquad | z_2 | \leq 1 \qquad\qquad (7.71)$$

Stability testing can thus involve separate tests for (7.70) and (7.71).

Checking for equation (7.70) has been carried out by Huang through replacement of $B(z_1,0)$ by another polynomial through a bi-linear transformation which has to be checked for its Hurwitz property (where all its zeros have negative real parts). Another test involves the Schur–Cohn matrix whose elements are simple functions of $B(z_1,0)$ and tested by examining the signs of the leading principle minors of the matrix. This test is used by Anderson and Jury to evolve an efficient but still time-consuming algorithm for both of these tests.

Many stabilisation design methods use an iterative method of differential correction which can be structured into a linear programming technique, i.e., using linear programming in an iterative fashion to solve a non-linear optimisation problem.

The approach taken by Dudgeon [32] is to develop a two-dimensional discrete Hilbert transform to calculate the analytic phase function from the magnitude-squared frequency response. This is similar to the discrete Hilbert transform used by Read and Treitel [28] to stabilise unstable filters and to the work of Pistor [29].

Dudgeon's filter design leads to non-finite arrays which require truncation before the design can proceed. This can introduce significant perturbations in the designed magnitude-squared frequency response which are undesirable. The designed filter is however always a stable realisation.

Pistor's paper proposes a criterion based on the relationship of the stability of recursive filters to the absolute summability of the cepstra operators. In a practical application this filter design is used in the enhancement of radioscintigraphic images [33].

This approach has also been taken by Cappellini [34] to produce a rapid design of a stable recursive filter. The design procedure commences with a given one-dimensional recursive filter having a squared-magnitude frequency response

$$H(\omega) = \frac{\sum_{n=0}^{N} a(n) \cos n\omega}{\sum_{p=0}^{M} b(p) \cos p\omega} \qquad (7.72)$$

This is transformed into a two-dimensional function by means of mapping due to McClellan, referred to previously, in order to realise a filter of the form

$$H(\omega_1,\omega_2) = \cfrac{\sum_{n=0}^{N-1} \sum_{m=0}^{N-1} a(n,m) \cos n\omega_1, \cos m\omega_2}{\sum_{p=0}^{M-1} \sum_{q=0}^{M-1} b(p,q) \cos p\omega_1, \cos q\omega_2} \qquad (7.73)$$

This will not necessarily be a stable realisation and Pistor's method is used to decompose the obtained unstable filter into four stable one-quadrant filters. These are acted upon by the image data in cascade to derive the filtered output.

The coefficient matrix of the denominator of equation (7.73) is decomposed into the convolution of four equal matrices in each quadrant, recursive in four different directions, by evaluating the cepstrum and by dividing it as an addition of four matrices. The four cepstra are then converted into the corresponding filters which are shown to be stable. A similar procedure can be carried out for the numerator matrix, a(n,m), or this may be left unchanged since it will not affect the stability of the filter.

This decomposition procedure is approximate and can give rise to errors due to aliasing and truncation so that some compromise must be reached to obtain an acceptable filter. An alternative method of optimum design by Bednar and Farmer [35] achieves some simplification in the derivation of filter coefficient arrays required in the difference equation but at the expense of requiring larger amounts of computation including a check for stability at each iteration of the algorithm.

A completely unrelated stabilisation procedure is proposed by Shanks[24] known as the 'double least-squares' technique. The basis of his design method is that of a planar least-squares inverse filter. In simple terms a given array C is assumed for which we would like to find an array P such that C convolved with P gives a unit pulse array U. Since this is not generally possible then P is chosen such that the sum of the squares of the elements of $U-G$ is minimised, where $G = C*P$. This array P is termed the planar least-squares inverse (PLSI) of C. The property of a PLSI employed by Shanks is conjectural but appears to be valid. This states that given an arbitrary real finite array C, any PLSI of C is minimum phase. A major result arising from this conjecture is that a filter $F(Z_1,Z_2) = 1/P(Z_1,Z_2)$ must be stable if P is a PLSI. However filters designed using Shanks's method will give a final filter frequency response which will only be an approximation and, as with Dudgeon's filter, non-negligible distortions of the desired amplitude spectrum often occur.

Finally we consider the conversion from one-dimensional filter design. A two-dimensional array can be considered as a superposition of sub-arrays each of which contains a single row and filtered using one-dimensional techniques. Each sub-array can be filtered, if necessary with different one-dimensional filters, to give the required characteristics for the overall system. Dudgeon proposes a scheme of this kind where a one-dimensional sequence is formed by concatenating the N columns of an input array. This sequence is then processed by a one-dimensional IIR filter. The output sequence is then used to form a two-dimensional array by taking the first N points as the first column, the next N points as the second column and so on. The complete system is linear but shift-varying and results in only certain impulse responses being capable of realisation with this method.

An improvement on this situation is described by Manry and Aggarwal [36] who, in effect, produce a mapping of equation (7.63) into a one-dimensional realisation. They show that whilst this results in a time-varying filter, the approximate performance to the desired two-dimensional filter can be obtained with a one-dimensional time-invariant filter under given conditions. There is a saving in computation time and memory using Aggarwal's method providing some error in grey level realisation is acceptable.

Shanks also gives a technique of two-dimensional design from the one-dimensional case by change of variable. This uses the well-known method of applying a bi-linear transformation to an analog filter in one-dimension. For the two dimensional case it is necessary first to project the one-dimensional analog filter into two dimensions. A given Laplace transform, $H(s_1)$, is equated to a two-dimensional analog form, $H(s_1, s_2)$, by means of a transformation

$$S_1 = S_1' \cos\theta + S_2' \sin\theta$$
$$S_2 = S_2' \cos\theta - S_1' \sin\theta \tag{7.74}$$

where θ represents an angle of rotation for the filter response in the two-dimensional Fourier plane. The bi-linear transformation is then applied to each variable

$$S_1' = \frac{\omega - 1}{\omega + 1} \qquad S_2' = \frac{Z - 1}{Z + 1} \tag{7.75}$$

to obtain a two-dimensional IIR filter. A variety of two-dimensional filter types can be realised by cascading filters having different angles of rotation and different one-dimensional prototype. A drawback with this technique is that a trial-and-error method may be needed to synthesise desired filter response and if the number of cascaded sections is high then it may be a very inefficient process.

References

1. Special issue on Digital image processing. *Computer,* **7**, 5, May 1974.
2. HUNT, B. R. Digital image processing. *Proc. IEEE,* **63**, 4, 693–706, April 1975.

3. ROSENFELD, A. Picture processing by computer. *Tech. Report*, 68−71, June 1968.
4. PROSSER, R. T. A multi-dimensional sampling theorem. *J. Maths. Anal. Appl.*, **16**, 574−84, 1966.
5. SWING, R. E. The optics of microdensitometry. *Opt. Eng.*, **12**, 185−98, November 1973.
6. CANNON, M. Blind deconvolution of spatially invariant image blurs with phase. *IEEE Trans. Acoustics, Speech and Signal Processing*, **ASSP-24**, 1, 58−63, February 1976.
7. HELSTROM, C. W. Image restoration by the method of least-squares. *J. Opt. Soc. America*, **57**, 3, 297−303, March 1967.
8. SLEPIAN, D. Restoration of photographs blurred by image motion. *Bell Syst. Tech J.*, **46**, 2353−63, December 1967.
9. SONDHI, M. M. Image restoration: The removal of spatially invariant degradations. *Proc. IEEE*, **60**, 842−53, July 1972.
10. CASTAN, S. Image enhancement and restoration. *NATO ASI on Digital Image Processing and Analysis*, Bonas, France, June 14−25, 1976.
11. HALL, E. L., *et al.* A survey of preprocessing and feature extraction techniques for radiographic images. *IEEE Trans. Computers*, **C-20**, 1032−44, September 1971.
12. ANDREWS, H. C. *Computer Techniques in Image Processing.* Academic Press, 1970.
13. UNDERWOOD, S. A., and AGGARWAL, J. K. Methods of edge detection in visual scenes. *Proc. IEEE Int. Symp. on Circuit Theory*, 45−51, 1973.
14. SHEPPARD, J. J. Pseudo-color as a means of image enhancement. *Am. J. Optom. Arch. Am. Acad. Optom.*, **46**, 735−54, October 1969.
15. ANDREWS, H. C., TESCHER, A. G., and KRUGER, R. P. Image processing by digital computer. *IEEE Spectrum*, 20−32, July 1972.
16. MEARS, R. W., WHITEHOUSE, H. J., and SPEISER, J. M. Television encoding using a hybrid discrete cosine transform and a differential pulse code modulator in real time (see *PRATT'S NATO ASI, Bonas*).
17. AHMED, N., NATARAJAN, T., and RAO, K. R. Discrete cosine transform. *IEEE Trans. Computers*, 90−3, January 1974.
18. FUSCO, L., and JEANRENAUD, A. A convolution program for digital filtering of images. *Proc. Florence Seminar on Digital Filtering*, Teorema, Firenze, Italy, September 1972.
19. HU, J. V., and RABINER, L. R. Design techniques for two-dimensional digital filters. *IEEE Trans. on Audio and Electroacoust.*, **Au-20**, 4, 249−57, October 1972.
20. HUANG, T. S. Two-dimensional windows. *IEEE Trans. Audio and Electroacoust.*, **Au-20**, 88−9, March 1972.
21. McCLELLAN, J. H. The design of two-dimensional digital filters by transformations. *Proc. 7th Princeton Conf. Information Science and Systems*, 247−51, 1973.
22. CALZINI, M., CAPPELLINI, V., AND EMILIANI, P. L. Alcuni filtri numerici bidimensionali con risposta impulsiva finita. *Alta Frequenza*, **XLIV**, 12, 747−53, 1975.
23. CAPPELLINI, A. G., CONSTANTINIDES, A. G. and EMILIANI, P. I.

Digital Filters and their Applications. Academic Press, London (in press).
24. SHANKS, J. L., TREITEL, S., and JUSTICE, H. H. Stability and synthesis of two-dimensional recursive filters. *IEEE Trans. Audio and Electroacoust.* **AU-20**, 5−128, June 1972.
25. HUANG, T. S. Stability of two-dimensional recursive filters. *IEEE Trans. Audio and Electroacoust.*, **AU-20**, 158−63, June 1972.
26. ANDERSON, B. D. O., and JURY, E. I. Stability of multi-dimensional digital filters. *IEEE Trans. Circuit and Syst.*, **CAS-21**, 300−4, March 1974.
27. MARIA, G. A., and FAHMY, M. M. On the stability of two-dimensional digital filters. *IEEE Trans. Audio and Electroacoust.*, **AU-21**, 470−2, October 1973.
28. READ, R. A., and TREITEL, S. The stabilisation of two-dimensional recursive filters via the discrete Hilbert transform. *IEEE Trans. Geoscience, Elec.*, **GE-11**, 153−60, July 1973.
29. PISTOR, P. Stability criteria for recursive filters. *IBM., J. Res. Dev.*, **18**, 59−71, January 1974.
30. DUDGEON, D. E. Using a two-dimensional discrete Hilbert transform to factor two-dimensional polynomials. *MIT Res. Lab. Electron. Quart. Prog. Rep.*, **113**, 143−50 April 1974.
31. MERSEREAU, R. M., and DUDGEON, D. E. Two-dimensional digital filtering. *Proc. IEEE,* **63**, 4, 610−23, April 1975.
32. DUDGEON, D. E. Two-dimensional recursive filter design using differential correction. *IEEE Trans. Acoustics, Speech and Sig. Proc.*, **ASSP-23**, 3, 264−7, June 1975.
33. PISTOR, P., Digital processing of scintigraphic images by two-dimensional recursive Wiener filters. *Report 70.12.006 IBM Deutschland Heidelberg Scientific Centre*, 1970.
34. BERNABO, M., CAPPELLINI, V., and EMILIANI, P. L., A method for designing two-dimensional recursive digital filters having a circular symmetry. *Florence Conf. on Digital Signal Proc.*, Univ. Florence, Italy, 196−203, September 1975.
35. BEDNAR, J. B., and FARMER, C., Implementation of Chebyshev digital filter design. *IEEE Trans. Circuits Syst.* (to be published).
36. MANRY, M. T., and AGGARWAL, J. K., Picture processing using one-dimensional implementations of discrete planar filters. *IEEE Trans. Acoustics, Speech and Sig. Proc.*, **ASSP-22**, 3, 164−73, June 1974.

AUTHOR INDEX

SUBJECT INDEX